Applications and Innovations in Intelligent Systems VIII

Springer-Verlag London Ltd.

Ann Macintosh, Mike Moulton
and Frans Coenen (Eds)

Applications and Innovations in Intelligent Systems VIII

Proceedings of ES2000, the Twentieth SGES International Conference on Knowledge Based Systems and Applied Artificial Intelligence, Cambridge, December 2000

Springer

Ann Macintosh, BSc, CEng
International Teledemocracy Centre, Napier University, Edinburgh, UK

Mike Moulton, BSc
Department of Accounting and Management Science, Portsmouth Business School, University of Portsmouth, Portsmouth, UK

Frans Coenen, PhD
Department of Computer Science, University of Liverpool, Liverpool, UK

ISBN 978-1-85233-402-4 ISBN 978-1-4471-0275-5 (eBook)
DOI 10.1007/978-1-4471-0275-5

British Library Cataloguing in Publication Data.
A catalog record for this book is available from the British Library.

Library of Congress Cataloging-in-Publication Data
A catalog record for this book is available from the Library of Congress.

Originally published by Springer-Verlag London Berlin Heidelberg in 2001

Typesetting: Camera ready by contributors

34/3830-543210 SPIN 10786357

APPLICATIONS PROGRAMME CHAIRMAN'S INTRODUCTION

Ann Macintosh
Napier University, UK

The papers in this volume are the refereed application papers presented at ES2000, the Twentieth SGES International Conference on Knowledge Based Systems and Applied Artificial Intelligence, held in Cambridge in December 2000. The scope of the Application papers has expanded over recent years to cover not just innovative applications using traditional knowledge based systems, but also to include applications demonstrating the whole range of AI technologies. This volume contains thirteen refereed papers describing deployed applications or emerging applications, together with an invited keynote paper by Dr. Daniel Clancy of NASA Ames Research Centre.

The papers were subject to refereeing by at least two "expert" referees. All papers which were controversial for some reason were discussed in depth by the Application Programme Committee. For the application stream, a paper is acceptable even if it describes a system which has not yet been installed, provided the application is original and the paper discusses the kinds of things that would help others needing to solve a similar problem. Papers have been selected to highlight critical areas of success (and failure) and to present the benefits and lessons learnt to other developers.

Papers this year cover topics as diverse as: KBS for maintaining offshore platforms; Data Mining to predict corporate business failure; integrated AI techniques to support field service engineers; Natural Language applied to the Data Protection Act; knowledge management and the application of neural networks.

This year the best refereed application paper is being presented for the innovative application of AI techniques in an unusual domain. The paper describes a system, called ASPERA, which uses natural language and case-based reasoning techniques to compose formal Spanish poetry.

This is the eighth volume in the *Applications and Innovations* series. The Technical Stream papers are published as a companion volume under the title *Research and Development in Intelligent Systems XVII.*

The conference was organised by SGES, the British Computer Society Specialist Group on Knowledge Based Systems and Applied Artificial Intelligence. On behalf of the conference organising committee I should like to thank all those who contributed to the organisation of this year's application programme, in particular the programme committee members, the referees and our administrators Linsay Turbert and Maria Doran.

Ann Macintosh
Applications Programme Chairman, ES2000

ACKNOWLEDGEMENTS

ES2000 CONFERENCE COMMITTEE

Dr Frans Coenen, University of Liverpool *(Conference Chairman)*
Dr Robert Milne, Intelligent Applications Ltd *(Deputy Conference Chairman, Finance and Publicity)*
Richard Ellis, Stratum Management Ltd *(Deputy Conference Chairman, Exhibition)*
Dr. Adrian Hopgood, Open University *(Tutorial Organiser)*
Ann Macintosh, Napier University *(Application Programme Chair)*
Mike Moulton, University of Portsmouth *(Deputy Application Programme Chair)*
Professor Max Bramer, University of Portsmouth *(Technical Programme Chair)*
Dr. Alun Preece, University of Aberdeen *(Deputy Technical Programme Chair)*

APPLICATION PROGRAMME COMMITTEE

Ann Macintosh, Napier University *(Chair)*
Mike Moulton, University of Portsmouth *(Deputy Chair)*
Richard Ellis, Stratum Management Ltd
Rick Magaldi, British Airways
Richard Wheeler, Starlab Research Laboratories

APPLICATION PROGRAMME REFEREES

Stuart Aitken, University of Edinburgh
Ruth Aylett, University of Salford,
Max Bramer, University of Portsmouth
Paul Chung, Loughborough University
Richard Ellis, Stratum Management Ltd
John Gordon, Applied Knowledge Research Institute
Peter Gray, University of Aberdeen
Robert de Hoog, University of Amsterdam
Ann Macintosh, Napier University
Rick Magaldi, British Airways
Rob Milne, Intelligent Applications Ltd
Alan Montgomery, InferMed
Mike Moulton, University of Portsmouth
Paul Slater, Compaq Computer Manufacturing Limited
Ian Watson, The University of Auckland
Richard Wheeler, Starlab Research Laboratories

CONTENTS

SESSION 3 (ENGINEERING APPLICATIONS)

APPLICATION KEYNOTE ADDRESS

Model-based System-level Health Management for Reusable Launch Vehicles

Daniel J. Clancy, PhD
Riacs Corp
NASA Ames Research Center
MS 269-2, Moffett Field, CA 94035
clancy@ptolemy.arc.nasa.gov

ABSTRACT

The next generation of reusable launch vehicles are expected to radically reduce the cost of accessing space thus enabling a broad range of endeavors including the commercialization of space and further manned exploration of the inner solar system. A reduction in operational costs requires more sophisticated techniques for monitoring and controlling the vehicle while in flight and techniques to streamline the ground processing of the vehicle. For both tasks, it is often necessary to synthesize information obtained from multiple subsystems to detect and isolate both hard failures and degraded component performance.

Traditionally, the synthesis of information from multiple components or subsystems is performed by skilled ground control and maintenance teams. In the future, much of this task will need to be performed by more sophisticated software systems that are able to reasoning about subsystem interactions. Performing this task using a traditional software programming paradigm is challenging due to the myriad of interactions that occur between subsystems especially when one of more components are performing in some off-nominal fashion. Model-based programming addresses this limitation by encoding a high-level qualitative model of the device being monitored. Using this model, the system is able to reason generatively about the expected behavior of the system under the current operating conditions, detect off-nominal behavior and search for alternative hypotheses that are consistent with the available observations. In this paper, we will talk about the Livingstone model-based health management system. Livingstone was initially developed and demonstrated as part of the Remote Agent Experiment on Deep Space One. Future flight experiments are planned for both the X-34 and the X-37 reusable launch vehicles.

Introduction

One of the critical challenges facing NASA as it looks toward the 21st century is developing technology to facilitate the cost effective and safe operation of the myriad of challenging missions that are planned. This problem is particularly critical when considering upgrades to the Space Shuttle. Each shuttle mission currently costs in the neighborhood of 400 million dollars and requires a ground control team of over 200 people to ensure the successful execution of the mission. One of the primary tasks performed by this team is managing the health of the shuttle to detect and isolate component failures and to select the appropriate recovery action to ensure the health and safety of the crew and when possible the success of the mission. Similarly, it is also critical to reduce the overall cost of processing and maintaining the systems in-between missions. Currently, these costs are often dominated by an exhaustive preventative maintenance strategy that requires many invasive maintenance procedures that increase the probability of a human error leading to a component failure.

One of the key challenges that must be addressed when managing the health of a complex physical system is the synthesis of information across interacting components and sub-systems. The system-level health management task is often performed by skilled ground control and maintenance teams. In the future, much of this task will need to be performed by more sophisticated software systems that are able to reasoning about subsystem interactions. Performing this task using a traditional software programming paradigm is challenging due to the myriad of interactions that occur between subsystems especially when one of more components are performing in some off-nominal fashion. Model-based programming addresses this limitation by encoding a high-level qualitative model of the device being monitored. Using this model, the system is able to reason generatively about the expected behavior of the system under the current operating conditions, detect off-nominal behavior and search for alternative hypotheses that are consistent with the available observations. In this paper, we will talk about the Livingstone model-based health management system. Livingstone was initially developed and demonstrated as part of the Remote Agent Experiment on Deep Space One. Future flight experiments are planned for both the X-34 and the X-37 reusable launch vehicles.

System-level Health Management

Assessing the health of a complex physical device is a difficult task that requires reasoning at multiple levels of detail. In its simplest form, the task can be viewed as a classification task in which the sensor data collected from the vehicle is classified as either in-family (i.e. nominal) or out-of-family (i.e. anomalous). Once this classification is performed, the task then must focus on the isolation of the source of the anomalous behavior. Once again, this can be viewed as a classification task in which the classes respond to all of the possible failures of the device. Within our approach, this classification task is performed using a generative model that can be used to predict the expected behavior of the device. Of course, the use of a model for such a task is nothing new. Models have been used for many years to try and characterize the behavior of a system. Many of the existing techniques, however, attempt to use a detailed quantitative model of a given component or sub-system. This model attempts to make precise predictions regarding how a given system will behave. The Linear Engine Model (LEM) developed by Rocketdyne of the Space Shuttle Main Engine (SSME) is a good example of this type of a model. In this example, a linearized model of the engine was developed that describes the nominal operating scenario along with 6 different potential failure modes. Developing such a model, however, is often quite difficult and expensive. Furthermore, the results provided by these approaches is often sensitive to inaccuracies that might exist in the model. Of course, a variety of techniques have been developed to address these an other problems and in general these techniques work well for providing a high-fidelity, detailed diagnosis of a complex component. The problem with these approaches is that they certainly do not scale to larger systems. For example, to develop a high-fidelity model of the entire space shuttle along with device failure modes is truly impossible due to inherent imprecision that would always exist in the model due to complex non-linearities within the system.

STS-93 Example

System-level health management is intended to address this problem by reasoning across components and sub-systems. Traditionally, this task is either performed by humans or by static procedural code that is designed to handle a set of pre-defined failure scenarios. To help better understand this problem, it is best to use an example that occurred in a recent Space Shuttle mission. In STS-93, a very slight liquid hydrogen leak developed shortly after lift-off. The current engine controller adjusts the liquid oxygen flow rate based upon the sensed flow rate of the liquid hydrogen to ensure that the mixture ration is correct. Unfortunately, the liquid hydrogen sensor was upstream from the point of the leak. As a result, the sensor detected an increase in the flow-rate of the liquid hydrogen causing the rate of LOX to be increased as well. This in turn caused the engine to run hot since the mixture ratio was off; however, the actual thrust generated by the engine was below normal. This eventually translated into an early main engine shut-off with the Shuttle ending up below the expected orbit. What makes this example particularly interesting is the fact that the discrepancy within any one of these readings was below the noise threshold and as a result it did not cause an alarm. As a result, the anomaly was not detected until the error accumulated and was manifested in the vehicle trajectory.

This example provides a good demonstration of the challenges that must be addressed in automating the system-level health management task. A traditional software development approach to this problem might have an engineer thinking through how the system might respond to the hypothesized failure, identifying the manifestation of the failure in the existing sensor suite, and writing static procedural code that identifies this failure when a specified signature results in the sensors. The key limitation of this approach is that the enumeration of all combinations of failures is impossible due to the breadth of possible faults and the possibilities of multiple failures occurring at once.

Instead, lets look at how a ground control team might detect and isolate this failure. If the failure had been considered before, then just as in the case of a procedural code the ground control team would have been trained to detect this failure based upon the expected signature in the sensor suite. Note, however, that in this case, detection and isolation of the failure requires the ground control team to look across subsystems. Thus, the engine team might notice that the engine is running a little hot while the navigation team might notice that the vehicle is a little off the expected trajectory. Individually, neither observation is sufficient for identifying and isolating the failure. In the case of the Shuttle, the back rooms would be responsible for reporting the detected anomaly to the front room which would in turn be responsible for synthesizing the information and making a diagnosis. If the failure was not one that the team had been extensively trained on, ground control personnel might *reason through* a number of hypothetical failures that could have resulted the observed anomalies. In performing this task, they would use an internal model of the device to *predict* what is expected. In this case the reasoning would go:

> *"If a slight leak occurred in the liquid hydrogen line, this would cause the LOX flow rate to increase, the engine to run hot, the thrust would be low and eventually we would be short of our expected trajectory."*

Since the observations matched this hypothesis, the ground control team could then select the appropriate recovery action to ensure the safety of the vehicle. In this particular instance, doing nothing was probably the appropriate response; however, it is easy to see how variations in this scenario might have required intervention by the ground control team.

Model-based health management refers to the achievement of robust operation through a growing set of reusable artificial intelligence (AI) reasoning algorithms that reason about first principles models of physical systems (e.g. spacecraft). In this context, a *model* is a logical or mathematical representation of a physical object or piece of software. A *first principles* model captures what is true about behavior or structure of the object (e.g. fluid flows through an open valve unless it is clogged). This is as opposed to traditional programs or rule-based expert systems, which capture what to do (e.g. turn on valves A, B, & C to start fuel flow) but unfortunately work only in certain implicit contexts (e.g. valves A, B, & C are working and A, B & C happen to control the fuel flow). Furthermore, the type of models used within this approach tend to be *qualitative* in nature. By this we mean that the specification used is an abstraction of more traditional quantitative models of a device. For example, as opposed to stating the precise quantitative relationship between two quantities, the model might simply identify the existence of a monotonic relationship between two related parameters. Furthermore, the values for these parameters are defined using a discrete state space such as *(low, medium, high).* The use of a qualitative model is one of the key benefits of this approach since it greatly reduces the cost of developing and validating the model. Furthermore, this approach is actually quite consistent with what is traditionally done by humans when performing this task. Thus, they might think of variables being *above normal* and *below normal* when reasoning at a system level.

Since model-based health management systems do not contain an encoding of what to do in each situation, they must reason about the appropriate action to take or conclusion to draw based upon their models and the currently available information about the environment. The past few decades of AI research have produced reasoning engines that can plan a course of action to achieve a goal, identify the current state of a physical system, reconfigure that system to enable some function (e.g. make the engine thrust) and so on from a first principles model.

Reasoning directly from the model, the current observations of the world and the task at hand provides many large advantages over traditional software development. Not least among these are that the system is robust in uncertain environments since it was not hard coded to respond to certain situations, the models and inference engines can be reused, and the models explicitly capture the assumptions about the system that are being relied upon to control it.

Benefits of Model-Based Health Management

A model-based health management system provides a number of benefits that are critical in the development of a robust health management architecture required for a next generation reusable launch vehicle. Note that while the previous discussion described the task of a ground control team, the objective of the technology being developed is to eventually enable the integration of this functionality into the real-time control loop that is on-board the vehicle. While various control techniques have been developed over the years, many of these techniques focus on the low-level control response required to maintain the system within a stable operating regime. The control techniques are then augmented with a higher-level discrete controller that is often implemented using a traditional software development methodology. Developing controllers in this manner is often time-consuming and often the resulting controllers are limited in their ability to handle novel component interactions that were not explicitly anticipated by the software developer. Furthermore, for devices that must operate without operator intervention for extended periods of time, it is often quite difficult if not impossible to write software that can handle all of the possible combinations of faults that can potentially occur over time.

A model-based controller addresses this problem by using a declarative specification of the device being controlled and the goals that are to be achieved. This results in increased safety and reliability while providing a significant decrease in the overall development costs due to code reuse and the compositional nature of the modeling paradigm. Finally, the use of a declarative specification simplifies the development of an advanced user interface that can be used to monitor the state of the device and to query the controller to obtain information about the actions that have been taken and the justification for these actions. This capability tends to decrease the overall system operation costs since the man power required to monitor the device is significantly reduced.

Livingstone

As mentioned previously, Livingstone is a model-based health management system that has been developed at NASA Ames Research Center (ARC). Its function is to infer the current state (mode) of each relevant device making up the system being controlled and to recommend actions that can reconfigure the system so that it achieves the currently desired configuration goals, if possible. In practice, these configuration goals could be provided by a human operating some apparatus by issuing high level configuration commands, or by some automated system such as the Smart Executive (Exec) mentioned above, which decomposes a high level plan into a series of configuration goals to be achieved. Purely for the sake of the discussion below, we will assume the Exec is providing the configuration goals and that the system being controlled is a spacecraft. Figure 1 provides a general Integrated Vehicle Health Management (IVHM) architecture that includes Livingstone long with a smart executive.

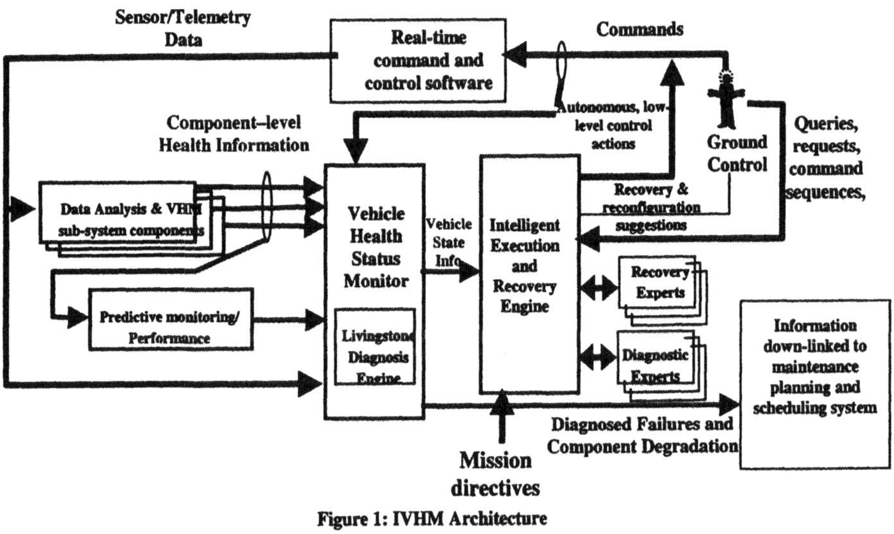

Figure 1: IVHM Architecture

To track the modes of system devices, Livingstone eavesdrops on commands that are sent to the spacecraft hardware. As each command is executed, Livingstone receives observations from spacecraft's sensors, abstracted by monitors in the real time control software for the Attitude Control Subsystem (ACS), communications bus, or whatever hardware is present. Livingstone combines these commands and observations with declarative models of the spacecraft components to determine the current state of the system and report it to the Exec. A pathologically simple example is shown schematically in Figure 2. In the nominal case, Livingstone merely confirms that the commands had the expected effect on spacecraft state. In case of failure, Livingstone diagnoses the failure and the current state of the spacecraft and provides a recovery recommendation. A single set of models and algorithms are exploited for command confirmation, diagnosis and recovery.

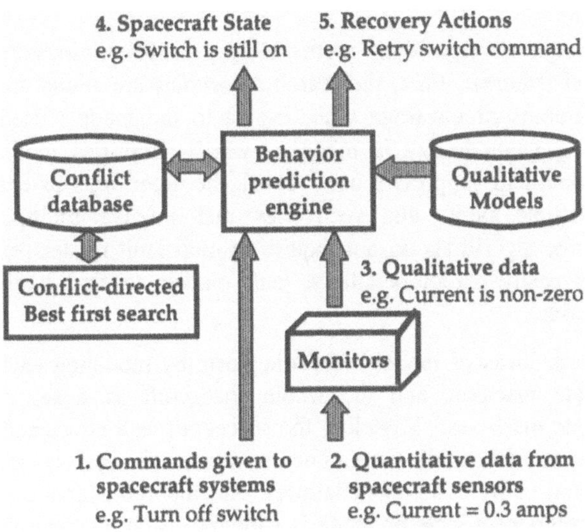

Figure 2. Information Flow in Livingstone

The capabilities of the Livingstone inference engine can be divided into two parts: mode identification (MI) and mode reconfiguration (MR). MI is responsible for identifying the current operating or failure mode of each component in the spacecraft. Following a component failure, MR is responsible for suggesting reconfiguration actions that restore the spacecraft to a configuration that achieves all current configuration goals currently required. Livingstone can be viewed as a discrete model-based controller in which MI provides the sensing component and MR provides the actuation component. MI's mode inference allows the Exec to reason about the state of the spacecraft in terms of component modes or even high level capabilities such as "able to produce thrust" rather than in terms of low level sensor values. MR supports the run-time generation of novel reconfiguration actions to return components to the desired mode or to re-enable high level capabilities such as "able to produce thrust".

Livingstone uses algorithms adapted from model-based diagnosis [1i, 1ii] to provide the above functions. The key idea underlying model-based diagnosis is that a combination of component modes is a possible description of the current state of the spacecraft only if the set of models associated with these modes is consistent with the observed sensor values. Following de Kleer and Williams [13], MI uses a conflict directed best-first search to find the most likely combination of component modes consistent with the observations. Analogously, MR uses the same search to find the least-cost combination of commands that achieve the desired goals in the next state. Furthermore, both MI and MR use the same system model to perform their function. The combination of a single search algorithm with a single model, and the process of exercising these through multiple uses, contributes significantly to the robustness of the complete system. Note that this methodology is independent of the actual set of available sensors and commands. Furthermore, it does not require that all aspects of the spacecraft state are directly observable, providing an elegant solution to the problem of limited observability.

The use of model-based diagnosis algorithms immediately provides Livingstone with a number of additional features. First, the search algorithms are sound and complete, providing a guarantee of coverage with respect to the models used. Second, the model building methodology is modular, which simplifies model construction and maintenance, and supports reuse. Third, the algorithms extend smoothly to handling multiple faults and recoveries that involve multiple commands. Fourth, while the algorithms do not require explicit fault models for each component, they can easily exploit available fault models to find likely failures and possible recoveries.

Livingstone extends the basic ideas of model-based diagnosis by modeling each component as a finite state machine, and the whole spacecraft as a set of concurrent, synchronous state machines. Modeling the spacecraft as a concurrent machine allows Livingstone to effectively track concurrent state changes caused either by deliberate command or by component failures. An important feature is that the behavior of each component state or mode is captured using abstract, or qualitative, models [14]. These models describe qualities of the spacecraft's structure or behavior without the detail needed for precise numerical prediction, making abstract models much easier to acquire and verify than quantitative engineering models. Examples of qualities captured are the power, data and hydraulic connectivity of spacecraft components and the directions in which each thruster provides torque. While such models cannot quantify how the spacecraft would perform with a failed thruster for example, they can be used to infer which thrusters are failed given only the signs of the errors in spacecraft orientation. Such inferences are robust since small changes in the underlying parameters do not affect the abstract behavior of the spacecraft. In addition, abstract models can be reduced to a set of clauses in propositional logic. This form allows behavior prediction to take place via unit propagation, a restricted and very efficient inference procedure.

It is important to note that the Livingstone models are not required to be explicit or complete with respect to the actual physical components. Often models do not explicitly represent the cause for a given behavior in terms of a component's physical structure. For example, there are numerous causes for a stuck switch: the driver has failed, excessive current has welded it shut, and so on. If the observable behavior and recovery for all causes of a stuck switch are the same, Livingstone need not closely model the physical structure responsible for these fine distinctions. Models are always incomplete in that they have an explicit unknown failure mode. Any component behavior that is inconsistent with all known nominal and failure modes is consistent with the unknown failure mode. In this way, Livingstone can still infer that a component has failed, though the failure was not foreseen or was simply left unmodeled because no recovery is possible. By modeling only to the level of detail required to make relevant distinctions in diagnosis (distinctions that prescribe different recoveries or different operation of the system) we can describe a system with qualitative "common-sense" models which are compact and quite easily written.

Applications

Researchers at NASA Ames are currently involved in a number of flight experiments and demonstrations of the use of the Livingstone system. These activities include IVHM flight experiments on both the X-34 and the X-37 experimental reusable launch vehicles, a collaboration with Johnson Space Center on autonomous control of closed-loop ecological life support systems and a collaboration with Kennedy Space center to autonomously control an In-situ Propellant Production (ISPP) facility for Mars. This section provides a brief overview of each of these activities.

X-34 Flight Experiment

The NASA IVHM Technology Experiment for X-vehicles (NITEX) has been selected to fly on the X-34 reusable launch vehicle being developed by Orbital Sciences Corporation. NITEX is a being developed collaboratively between NASA Ames Research Center, NASA Glenn Research Center, and Kennedy Space Center. The goal of the X-34 IVHM flight experiment is to advance the technology readiness levels of selected IVHM technologies within a flight environment and to begin the transition for these technologies from experiment status into accepted RLV baseline designs. Multiple flights of the experiment are planned beginning in April of 2001.

The experiment will monitor the X-34 vehicle throughout all mission phases using detailed diagnostic algorithms to detect degraded component performance as well as a system-level health monitoring system that integrates information from multiple components to perform real-time fault detection, isolation and recovery. In addition, the experiment will demonstrate the use of an advanced, user friendly ground station that combines information provided by the on-board IVHM software with information obtained while the vehicle was on the ground to provide high-level status information on the health of the vehicle along with the ability to access more detailed information when required. The ground station will also provide justification for the inferences made by the IVHM system and alternative recovery recommendations following a failure while in flight. The focus of the experiment will be on the X-34's Main Propulsion Subsystem (MPS) including the Fastrac Engine and the Reaction Control System (RCS). The primary component of this experiment as well as the other demonstrations described below is the Livingstone inference engine.

X-37 Flight Experiment

A similar flight experiment is also being developed to fly on Boeing's X-37 reusable launch vehicle. The X-37 vehicle will be deployed from the Space Shuttle, orbit for 21 days, and then perform an autonomous entry, descent and landing. The IVHM experiment will monitor the electrical power subsystem along with the electro-mechanical actuators (EMAs) throughout the flight. For this experiment, the software will be hosted on Boeing's Vehicle Management Computer (VMC). As a result, there are very tight real-time requirements with limited computational resources. Currently, the experiment expects to have only about 1.5% of the total CPU cycles. Thus, this experiment focuses on validating the Livingstone system on an actual flight computer with hard real-time constraints. Demonstrating the abilities of the technology within this highly constrained environment is critical to the eventual adoption of this technology.

Closed-Loop Ecological Life Support Systems (CELSS)

In order to transport and support humans for Mars expeditions, NASA's Human Exploration and Development of Space (HEDS) requirements state a need for autonomous operation of life support, ISRU and transport equipment. During a Mars expedition, autonomous plant operations would allow unmanned systems to prepare for human arrival, protect crew and resources by rapidly responding to critical failures, and free humans from routine operations, allowing greater exploration.

At NASA's Johnson Space Center (JSC), a closed loop life support testbed called Bioplex has been constructed. The Bioplex consists of three sections: a three story cylindrical living quarters similar to the Mars habitats discussed in various mission proposals; a plant chamber where wheat is grown to provide food and exchange CO_2 for O_2; and an incinerator chamber used to eliminate solid waste and produce CO_2. The most recent Bioplex testing is referred to as the Product Gas Transfer phase as it concentrates on generation and distribution of product gases (CO_2 from the crew and incinerator and O_2 from the plants) and does not yet address issues such as waste water recycling or power management.

A JSC advanced development group has developed an autonomous control system to operate the product gas transfer phase of Bioplex [15]. This system, based upon the 3T autonomy architecture [16], maintains the appropriate atmosphere in each chamber by extracting and storing product gases and coordinating activities such as firing the incinerator or opening the plant chamber for human access. The system successfully controlled gas transfer during test in which a human crew inhabited the

Bioplex for ninety days. It was not expected, however, to maintain operation in the face of failures, though many would likely occur over a 4-year mission.

We are currently working to integrate the Livingstone mode identification and reconfiguration engine with JSC's 3T architecture, adding to it the ability to determine the current state of the testbed and respond to anomalous situations or failures by performing high level, system-wide reasoning. This will result in a single, reusable architecture which maintains the best possible operation of a regenerative life support system and other complex physical plants during both nominal operation and failures, somewhat analogous to the autonomic and immune functions of a living organism.

We intend to demonstrate the combined system by maintaining operation of the testbed over an extended test period and providing both fully autonomous and human-centered operation. To test the system, an outside examiner will be employed to introduce failures into the testbed as desired which the system will diagnose and attempt to mitigate.

The second goal is to demonstrate and extend the ability of model-based systems to reduce analysis, development and operations costs. The testbed application will be rapidly developed with tools that could be used to develop mission applications. Users will develop and operate the testbed by manipulating explicit models with visual tools. If previous experience is to be believed, far less effort will be required to develop, understand and revise the system than in an approach where system model is implicit but still must be maintained.

If successful, this demonstration will increase the likelihood that autonomy technologies being developed by NASA are appropriate and sufficiently mature when they are required for HEDS missions to Mars and other destinations. It will also ensure that the necessary technologies can be integrated and will identify needed extensions before such shortcomings could impact the critical path of a mission. In addition, JSC will have a prototype of a reusable, fault-tolerant, high-capability autonomous control system and the expertise to apply this system to a flight experiment or mission. This could be applied to any complex physical system that must be controlled and maintained over an extended period of time such as spacecraft, power plants, ISRU machinery, and autonomous or semiautonomous surface vehicles.

In-situ Resource Utilization

In-situ resource utilization, or "living off the land", is critical to making a piloted Mars mission robust and affordable [1]. More specifically, it is envisioned that in-situ propellant production (ISPP) plants will arrive on Mars years before humans and begin combining hydrogen brought from Earth with CO_2 from the Martian atmosphere to create methane. This fuel will power the ascent vehicle that will lift the crew off Mars to begin their trip home in addition to powering any methane-fueled surface vehicles the astronauts might possess.

Though the chemical reactions involved are conceptually quite simple, on Mars they are somewhat complicated by issues such as the low atmospheric pressure and slow contamination of the ISPP catalysts by trace elements in the Martian atmosphere. To ensure that adequate ISPP capability is available for future Mars missions, NASA has begun to explore ISPP designs and build prototype hardware for operation in Mars-like test chambers. Both JSC and NASA Kennedy Space Center (KSC) are involved in early ISPP development, and the KSC team is integrating Livingstone into their ISPP prototyping efforts.

The short-term focus of this collaboration is to integrate Livingstone's ability to diagnose and mitigate failures with existing KSC model-based technology to gain experience with a model-based monitoring, diagnosis and recovery system for ISPP. A secondary short-term goal is to determine if any other autonomy technology previously invested in by NASA, for example the Smart Executive, can be reused on the ISPP testbed, thus increasing capability without greatly increasing cost.

A longer term goal is to continue research into control of physical systems which must continuously adjust their operation to unforeseen degradation in capability (for example an ISPP unit where Martian dust covers solar panels or slowly clogs air filters) rather than taking a discrete recovery action as Livingstone does. Related issues include reasoning about hybrid discrete/continuous systems, predictive diagnosis and relearning models of the continuous dynamic behavior of the system. This research should contribute to development of ISPP and other systems that are robust and yet run at the ragged edge of optimality throughout their lifetimes, neither being overly conservative nor exceeding their remaining degraded capabilities.

Acknowledgements

This paper touches on the work of a great many people too numerous to name here. The Autonomous Systems Group at NASA Ames Research Center consists of about twenty computer science researchers pursuing all manners of autonomy research, much of which was not mentioned here. Members of the JPL New Millennium Program and AI Group contributed to the Remote Agent architecture and to making it work on a flight platform. Advanced development groups at NASA GRC, NASA JSC (3T and Bioplex PGT), NASA KSC and JPL have shared their expertise with us and are helping to push the model-based autonomy technologies described here forward. In particular, I would like to acknowledge the members of the X-34 and X-37 IVHM experiment teams for their dedication to the advancement and demonstration of this technology.

References

Many of the following papers may be found on the World Wide Web at http://ic-www.arc.nasa.gov/ic/projects/mba/

[1] R. Zubrin and R. Wagner. The case for Mars: The plan to settle the Red Planet and why we must. The Free Press, 1996.

[2] S. J. Hoffman and D. I. Kaplan, editors. Human Exploration of Mars: The Reference Mission of the NASA Mars Exploration Study Team. NASA Special Publication 6107. July 1997.

[3] A. H. Mishkin, J. C. Morrison, T. T. Nguyen, H. W. Stone, B. K. Cooper and B. H. Wilcox. Experiences with operations and autonomy of the Mars Pathfinder microrover. In Proceedings of the IEEE Aerospace Conference, Snowmass, CO 1998.

[4] G. M. Brown, D. E. Bernard and R. D. Rasmussen. Attitude and articulation control for the Cassini Spacecraft: A fault tolerance overview. In 14th AIAA/IEEE Digital Avionics Systems Conference, Cambridge, MA, November 1995.

[5] B. C. Williams and P. Nayak, A Model-based Approach to Reactive Self-Configuring Systems, Proceedings of AAAI-96, 1996.

[6] B C. Williams and B. Millar. 1996. Automated Decomposition of Model-based Learning Problems. In Proceedings of QR-96.

[7] N. Muscettola, B. Smith, C. Fry, S. Chien, K. Rajan , G Rabideau and D. Yan, Onboard Planning For New Millenium Deep Space One Autonomy, Proceedings of IEEE Aerospace Conference, 1997

[8] B. Pell, E. Gat, R. Keesing, N. Muscettola, and B. Smith. Robust periodic planning and execution for autonomous spacecraft.

[9] B. Pell, D. E. Bernard, S. A. Chien, E. Gat, N. Muscettola, P. P. Nayak, M. D. Wagner, and B. C. Williams, An Autonomous Spacecraft Agent Prototype, Proceedings of the First International Conference on Autonomous Agents, 1997.

[10] D. E. Bernard et Al. Design of the Remote Agent Experiment for Spacecraft Autonomy. Proceedings of IEEE Aero-98.

[11] J. de Kleer and B. C. Williams, Diagnosing Multiple Faults, Artificial Intelligence, Vol 32, Number 1, 1987.

[12] J. de Kleer and B. C. Williams, Diagnosis With Behavioral Modes, Proceedings of IJCAI-89, 1989.

[13] J. de Kleer and B. C. Williams, *Artificial Intelligence*, Volume 51, Elsevier, 1991.

[14] . S. Weld and J. de Kleer, *Readings in Qualitative Reasoning About Physical Systems*, Morgan Kaufmann Publishers, Inc., San Mateo, California, 1990.

[15] D.Schreckenghost, M. Edeen, R. P. Bonasso, and J. Erickson. Intelligent control of product gas transfer for air revitalization.. Abstract submitted for 28th International Conference on Environmental Systems (ICES), July 1998.

[16] R. P. Bonasso , R.J. Firby, E. Gat, D. Kortenkamp, D. Miller and M. Slack. Experiences with an architecture for intelligent, reactive agents. In Journal of Experimental and Theoretical AI, 1997.

[17] J. Bresina, G. A. Dorais, K. Golden, D. E. Smith, R. Washington, Autonomous Rovers for Human Exploration of Mars. Proceedings of the First Annual Mars Society Conference. Boulder, CO, August 1998. To Appear.

[18] B. C. Williams and P. P. Nayak. Immobile Robots: AI in the New Millennium. In AI Magazine, Fall 1996.

[19] B. C. Williams and P. P. Nayak. A Reactive Planner for a Model-based Executive. In Proceedings of IJCAI-97.

[20] N. Muscettola. HSTS: Integrating planning and scheduling. In Mark Fox and Monte Zweben, editors, Intelligent Scheduling. Morgan Kaufmann, 1994.

[21] V. Gupta, R. Jagadeesan, V. Saraswat. Computing with Continuous Change. Science of Computer Programming, 1997.

BEST REFEREED APPLICATION PAPER

An Expert System for the Composition of Formal Spanish Poetry

Pablo Gervás
Universidad Europea - CEES, Madrid
pg2@dinar.esi.uem.es
www.uem.esi.es/~pg2

Abstract: The present paper presents an application that composes formal poetry in Spanish in a semiautomatic interactive fashion. ASPERA is a forward reasoning rule-based system that obtains from the user basic style parameters and an intended message; applies a knowledge-based preprocessor to select the most appropriate metric structure for the user's wishes; and, by intelligent adaptation of selected examples from a corpus of verses, carries out a prose-to-poetry translation of the given message. In the composition process, ASPERA combines natural language generation and CBR techniques to apply a set of construction heuristics obtained from formal literature on Spanish poetry. If the user validates the poem draft presented by the system, the resulting verses are analysed and incorporated into the system data files.

1. Introduction

The automatic generation of text is a well established and promising problem in AI, with numerous practical applications waiting in the sidelines for efficient and acceptable solutions. Existing systems have shown reasonable results in restricted domains [3,7,9], opening the way to considering how more elaborate texts - from the point of view of aesthetics and reader satisfaction - can be obtained [2,4,8]. The composition of poetry ranks among the most challenging problems of language generation, and is therefore a good testbed for techniques designed to improve the quality of generated texts. ASPERA (Automatic Spanish Poetry Expert and Rewriting Application) is a prose-to-poetry semiautomatic translator. By ingenious use of well accepted AI techniques (Natural Language Processing, Case Based Reasoning, Knowledge Based Systems), the application obtains from the user a prose description of the intended message and a rough specification of the type of poem required (length, mood, topic); selects appropriate metre and stanza (by resorting to a knowledge base on literary style); generates a draft of the poem (by applying CBR techniques to a database of previous poems); requests modification or validation of the draft by the user; and updates its own database of information (using NLP techniques to extract all the required linguistic information from the validated poem). ASPERA is designed to be used as a teaching aid for specific information technology subjects of the degree on Translation and Interpretation at the Universidad Europea - CEES.

2. Problem Description

On first acquaintance, the generation of poetry involves advanced linguistics skills and common sense, two of the major challenges that face AI in general. On the positive side, poetry has the advantage of not requiring exaggerate precision. To a certain extent, imposing restrictions over the form of a poem implies a slight relaxation on the specification of the content. Under this assumption, the general problem becomes tractable. There are three main challenges to be faced:

1. a specification of the formal requirements that define a correct poem under classical literary rules (in a format that can be used to drive the construction process)
2. appropriate management of an extensive vocabulary (the choice of words plays an important role in determining the quality of a poem)
3. correct combination of words in the poem to match both the intended message and the chosen metric structure

2.1 The Formal Rules of Spanish Poetry

The fact that the application was to be developed in Spanish presents important advantages over other languages. The phonetics of Spanish are quite straightforward to obtain from the written word. Most letters in Spanish sound the same wherever they appear in a piece of text, so the metrics, or the syllabic division, of a verse can be worked out algorithmically [11]. Spanish scholars have a love for rules, and there is a good set of formal rules [10] describing the conditions that a poem must fulfil in order to be acceptable.

The challenge becomes a simple problem of transforming the given evaluation rules (designed to be applied to an existing poem in order to ascertain its acceptability) into the corresponding construction rules.

Given that words are divided into syllables and each word has a unique syllable that carries the prosodic stress, the constraints that the rules have to account for are the following:

1. Specific strophic forms require different number of syllables to a line
2. Syllables may be run together under specific conditions - this is a form of poetic license called *synaloepha* - thereby shortening the syllable count of the line involved
3. The position of the stressed syllable of the last word of a line affects the metric count for that line (adding or subtracting one syllable to the count)
4. Not all possible stress patterns for a line are valid (depending on the length)

5. The rhyme of a word is made up of the last half of its stressed syllable and all following syllables to the end of the word, and each strophic form requires a different rhyming pattern

A poem may be an unstructured sequence of lines, but for the purpose of this application only poems of regular strophic forms are considered. For the purposes of the present application, the following strophic forms are relevant:

1. *romances*, a stanza of several lines of eight syllables where all even numbered lines rhyme together
2. *cuartetos*, a stanza of four lines of eleven syllables where the two outer lines rhyme together and the two inner lines rhyme together
3. *tercetos encadenados*, a longer poem made up of stanzas of three lines of eleven syllables linked together by their rhyme in a simple chain pattern ABA BCB CDC...

These three types have been chosen because each shows a different structural characteristic.

Type 1 presents a recurring rhyme that flows all along the poem. It uses only one rhyme, so many words that rhyme together are required for an acceptable result (our starting data have proven to be poor choices in this respect).

Type 2 presents a very simple but rigid structure. It stands for the simplest possible stanza with enough complexity to be distinguishable from prose (the simplification employed with respect to syntax/semantics makes it difficult for shorter poems to sound acceptable).

Type 3 presents a simple structure that recurs throughout the poem, but with the rhyme changing slowly as it moves down.

A constructive implementation of these rules was developed for the WASP system [4]. WASP was a forward reasoning rule-based system that used a set of construction heuristics obtained from these constraints to produce a poem from a set of words and a set of line patterns provided by the user. The system followed a generate and test method by randomly producing word sequences that met the formal requirements. Output results were impeccable from the point of view of formal metrics, but they were clumsy from a linguistic point of view and made little sense. An improved version of the construction strategies developed in WASP is the starting point of the generating module of ASPERA.

2.2 Guiding Word Choice

The set of words available as building blocks for the composition process play a crucial role in determining the quality of the results. Too narrow a choice of vocabulary can lead to formal requirements not being met. Too wide a choice can seriously reduce the efficiency of the system, leading to lengthy searches over vast amounts of words. A trade off must be found: enough words to compose the poem must be available, but they must all be words with a reasonable chance of finding their way into the final result. This is a part of the problem where intelligent solutions have a good chance of outperforming purely computational techniques.

The ASPID system [5] provided specific algorithms for the selection of a working set of words from an initial vocabulary using methods based on similarity calculations between the message proposed by the user for his poem and a corpus of already validated verses. Based on the similarity calculations, the system established a set of priorities over the complete available vocabulary. The next word to be added to the poem draft was initially looked for only among words marked with the highest priority, with the search extending in subsequent steps to words of lower priority only if none had been found in the previous step. This procedure improved search times considerably and it made possible computations with wider vocabulary coverage and narrower constraints on strophic forms. However, above a certain threshold (of vocabulary size and/or number of constraints imposed on the poem) even the method of establishing a priority ordering on the available words failed to ensure successful termination. The ASPID method of priority assignment is retained in ASPERA, but a prior step of knowledge-based pre-selection of the vocabulary based on user requirements has been added.

2.3 Fitting Words to Message and Metric Structure

There are several restrictions on the actual process of composition that must be observed.

The length of a poem is given by the number of lines (and the length of the lines) in the chosen strophic form. The intended message must be shortened or extended to adjust it to the length of the poem.

The basic unit for poem composition is not the sentence but the line. A poem may contain one or several sentences, but it is in its subdivision into lines that the constraints (position of stressed syllables, and rhyme) are imposed. A step of planning is required to distribute the contents of the intended message over the required number of lines.

Words at the end of lines may have additional constraints imposed on them (rhyme) by the chosen strophic form. These restrictions must be taken into account when planning the poem.

The words in a poem must be combined according to the syntax of the language in question, and must make sense according to their semantics. There are two alternative ways of achieving this: to provide the system with adequately rich lexicon, syntax and semantics for the language involved (as done in [8]) for English poetry), or to develop engineering solutions that achieve equivalent results without attempting to model the imposing complexity of the human language faculty. The former solution requires powerful tools capable of representing and manipulating the syntax and semantics of language. For the present application, the latter approach is preferred.

ASPERA resorts to a radical simplification of the linguistic skills underlying poem composition. The exhaustive knowledge approach is abandoned in favour of a heuristic engineering solution. Only the barest outline of a grammatical outline is provided (in the form of a line pattern) to ensure syntactic correctness. Semantic correctness is not enforced strictly, rather an approximate result is obtained by intelligent pre-selection of the choice of words together with a memory based approach to word combination.

The system is provided with a corpus of already validated verses. A Case Based Reasoning approach [1] is applied to this corpus in order to generate new verses. The words in these verses are marked with their corresponding part-of-speech (POS) tag. A line pattern is generated from these POS tags. Each line pattern is a set of tags, and each tag acts as place keeper for a possible word of the line. The tag is actually a string that represents information about part of speech, number, and gender of the word that would stand in that particular place in the pattern. Patterns act as seed for lines, therefore a pattern determines the number of words in a line, the particular fragment of sentence that makes up the line, and the set of words that can be considered as candidates for the line. By following this heuristic shortcut, the system is able to generate verses with no knowledge about grammar or meaning.

3. ASPERA: Automatic Spanish Poetry Expert - a Rewriting Application

ASPERA is a forward reasoning rule-based system that performs the following sequence of operations:

1. interacts with the user to obtain a specification of the desired poem (intended message, mood, setting);
2. searches its knowledge base to find the most appropriate strophic form to match that specification;
3. plans a poem draft by distributing the intended message over the chosen strophic form;
4. pre-selects and loads a task-specific vocabulary and corpus of verse examples for each fragment of the poem by adequately combining its data files (CBR Retrieve step I);
5. generates each of the lines of the poem draft by mirroring the POS structure of one (CBR Retrieve step II) of the pre-selected line examples combined with the words in the pre-selected vocabulary (CBR Reuse step);
6. presents the draft to be validated or corrected by the user (CBR Revise step); and
7. carries out a linguistic analysis of any validated poems in order to add the corresponding information to its data files to be used in subsequent computations (CBR Retain step).

ASPERA is written in CLIPS, a rule-based system shell developed by NASA. Earlier implementations were attempted using Prolog, but the nature of the line generation part of the problem, being a constructive combination of a set of elementary ingredients towards the achievement of a single whole which is the final poem, lends itself more easily to a forward-reasoning mode of operation. The system's poetry expert knowledge base had originally been coded in Prolog, but translation onto the new paradigm was presented no special problems.

3.1 Collection of Poem Specifications

The system interacts with the user to obtain specifications of the desired poem. The user is first asked a set of questions designed to ascertain a few elementary facts about his intentions. These will be used by the knowledge-based module of the system to select the right vocabulary and an adequate set of verse examples from the available corpus.

At present, the system is designed to operate with the following parameters:

- *approximate length of the poem*: The user has to provide a numerical value for the number of lines he would like to obtain. The system uses this number in working out an adequate strophic form (or combination of strophic forms) for the poem. The user is also asked whether the stated length is a rigid or flexible constraint.
- *rhyme structure*: Some strophic forms have rigid rhyme structures (there are constraints on the rhyme of every line) and others are more flexible (there are constraints on only some of the lines).
- *degree of formality*: Certain strophic forms are more formal than others. This distinction plays an important role in selecting the appropriate form.
- *setting*: The system provides a choice between an urban setting or a rural setting. This is a very restricted choice, but it may be extended at later stages.
- *mood*: As a first approximation, mood is interpreted either as positive or negative.

Length of poem and degree of formality are used specially to determine the strophic form. Setting and mood are used to select the correct vocabulary.

Once the basic parameters have been set, the user is asked to provide a prose paraphrase of this intended message. This paraphrase need not be grammatically sound, and may range from a full explicit text to a set of keywords that the user would like to show up in the poem. The intended message is stored as a list of words in a format amenable for later use by the system:

```
(message Peter loves Mary they go together to the beach)
```

Every word in the message is considered a good candidate to contribute to the final poem. Statistical methods of natural language processing applied in the field of information retrieval favour an initial trimming of user queries with a view to retaining for processing only terms that have relevance for the retrieval task. Stop lists are applied in order to eliminate empty words (pronouns, articles, prepositions...). For the present purposes, however, the presence of words of these types allows the user to control the style and appearance of the final result even more (or at least as much as) nouns, verbs and adverbs.

3.2 The Poetry Expert Knowledge Base

The Poetry Expert knowledge base contains two distinct sets of rules that encode specific knowledge concerning the relationship between strophic forms, the user's wishes, and the available data file. Their job is to select the best available combination of strophic form and data files to suit the user's request.

The first set of rules concerns the choice of strophic form and the selection of data files with the required sets of verse examples/patterns. They have been obtained by a knowledge acquisition process performed systematically over the author's personal experience, careful analysis of classical Spanish poetry, and available references on poetic analysis [10]. This gave rise to a complex set of rules that relate possible combinations of input parameters with possible outputs (strophic forms or combinations of strophic forms). The relationship is mostly associative, but requires a step of arithmetic calculation when processing the required length. The decisions embodied in the rules follow roughly the following guidelines:

- formal poem → strophic forms of 11 syllables.
- informal → strophic forms of 8 syllables.
- long and flexible → *romance* or *terceto encadenado*
- short and concise → *terceto*
- fully rhymed → *cuarteto* or *terceto encadenado*
- loosely rhymed → *romance* (long) or *terceto* (short)
- many similar rhymes → dos *cuartetos* (not too long) or *romance* (long)

The second set of rules associates the information obtained from the user regarding setting and mood desired for the poem with the various data files over which the available vocabulary is distributed.

The application of the Poetry Expert knowledge base results in a message to the user proposing a specific strophic form. If the user accepts the proposal, the corresponding data files are loaded and the system moves onto the next stage. If the user rejects the proposal, the system allows the users choice of strophic form to override its own suggestion. The knowledge base is then addressed for the right combination of data files and these are loaded. Mismatching specifications of poem length or degree of formality and choice of strophic form may lead to non termination.

3.3 The Elementary Planning Step

Given the user's choice of basic parameters the system uploads an appropriate combination of data files containing vocabulary extracts. A vocabulary entry for a word is a CLIPS fact of the form:

```
(word (cual desde_n) (numsil 2) (accent 2)
      (emp 0) (term 0) (cat NCMS)
      (rima en) (level nil))
```

An entry must provide all the necessary information for imposing the metric restrictions: length of the word in syllables (numsil), position of stresses syllable (accent), whether the word starts (emp) or ends (term) with a vowel, the POS tag for the word (cat), and the rhyme (rima). An additional level field is provided to enable priority marking of the words during later processing.

For the resulting choice of strophic form specific line patterns /examples are loaded. A line pattern/example is a CLIPS fact of the form:

```
((sample (n 72)
        (patt NCMS PDEL NCMS NCMS DET PPO3FS NCFS)
        (words desde_n del cielo error de la ventura)
        (beg 0) (end 1) (level nil))
```

where n is a unique identifier for the specific sample, patt is the set of POS tag corresponding to the sample - which acts as line pattern -, words is the actual line example, beg encodes information about whether such line starts a sentence, end encodes information about whether the line ends a sentence, and the level field is used for priority settings.

The words in the intended message provided by the user are distributed into as many fragments as there are lines in the chosen strophic form, retaining within each fragment the original order of appearance. For instance, supposing the message presented above were to be rewritten as a terceto, it would be split by the system into the following facts:

```
(fragment 1 Peter loves Mary)
(fragment 2  they go together)
(fragment 3 to the beach)
```

This is not considered a draft of the poem, but only as an approximate distribution of information over lines of the poem.

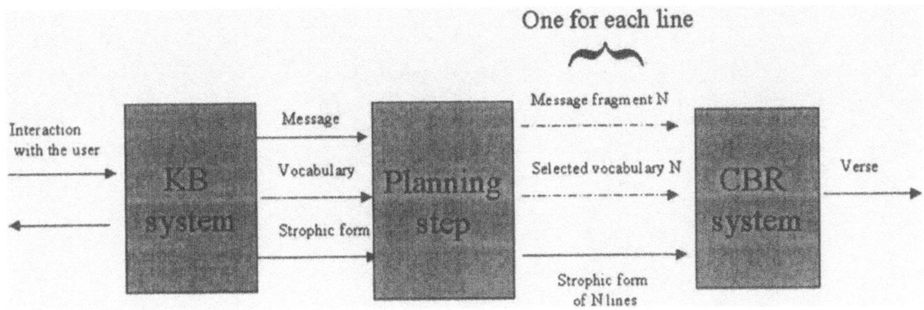

Figure 1 Basic structure of the ASPERA system

A module of the system works out the similarities between the line fragments and the available line examples/patterns. Each fragment/pattern pair is indexed with a numerical value indicating their similarity. In the initial approximation, this similarity value is worked out as the number of POS tags they have in common. The results of these similarity calculations are later used during the CBR Retrieve step.

Another module counts the number of rhymes available (if any) for each of the pre-selected words. These numbers are later used to assign specific words and patterns to each fragment during the CBR Retrieve step: a pattern is only good for a given line in the poem if there is a word of the necessary rhyme whose POS matches the last POS tag in the pattern.

The CBR Retrieve step can be considered to start during the planning stage, since line patterns/examples are assigned to the different fragments of the poem at this stage. This is because the retrieval of resources for one line is not independent from the assignation of resources to its neighbours. It is important, for instance, to ensure that consecutive lines are assigned sets of patterns that allow a correct linguistic fit between the resulting lines (the sentence fragment in the first line can be continued with the sentence in the following one, or there are patterns for starting a sentence on the following line if patterns in the first line include an end of sentence at the end of the line).

3.4 The CBR Approach to Line Generation

A different CBR process is set in motion for each line in the poem. This is done sequentially starting from the first line, but none of the different CBR processes progresses onto its next stage until all the other processes are ready to do so as well: all processes retrieve their vocabularies and line examples, all processes adapt their selection to generate a new line (the revision and analysis phases can be done in any order). This ensures that no two lines in the same poem are built following the same pattern and that words are not repeated in a poem.

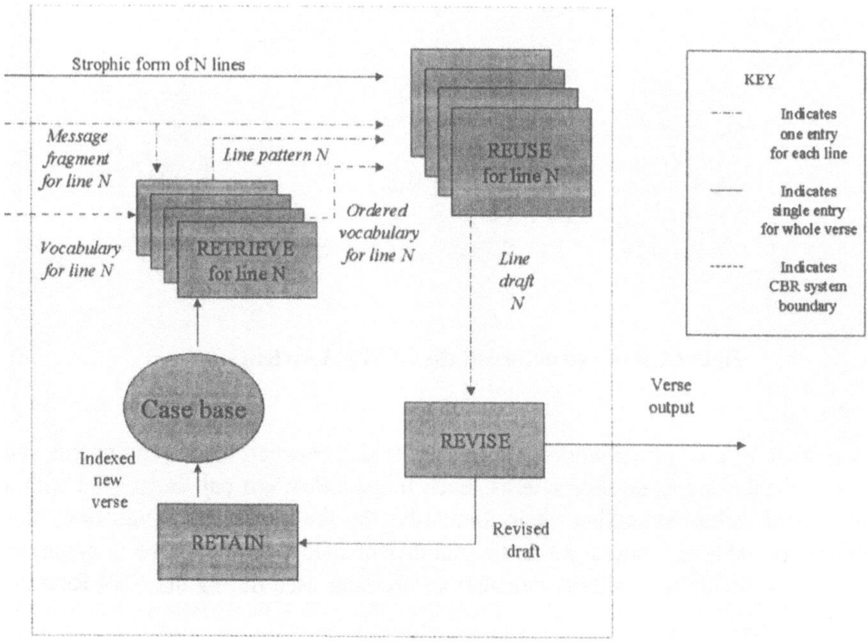

Figure 2 Verse generation using CBR

3.4.1 Retrieve step

For each fragment of the poem, adequate words and adequate line patterns/examples must be selected. Line examples/patterns have already been assigned during the planning stage.

A module assigns priorities to all the available words, so that only the most adequate words are considered for each line. The criteria used in this assignation are, in decreasing order of priority: words actually appearing in the corresponding fragment of the intended message, words with the required rhyme for the line.

These priorities can be modified once the initial lines have been generated, to include any words in the fragment of the intended message corresponding to previous lines that have not been used in the generation of the actual verse.

3.4.2 Reuse step

For each set of fragment of the intended message, set of line examples /patterns, and set of selected words, the basic generation algorithm following the constructive rules for the appropriate line length is applied.

The elementary generation units are the *line pattern* being followed (which acts as guiding form) and the *draft of the current line*. At each step of the generation the words are chosen to match the next POS tag in the line pattern (always according to their established priority ordering). If the chosen word meets the constructive requirements, it is appended to the draft of the current line and the process is iterated. The constructive requirements are implemented in the form of imperative Boolean functions.

If no valid extension is found, the system allows limited backtracking: the last word to be added can be removed and an alternative attempted. An additional process of annotation takes place to avoid a repetition of alternatives that have already been tried.

Additionally, the assignation of line examples/patterns to each fragment of the poem can be revised during the Reuse step if no solution is found with the existing assignation once all possible backtracking alternatives over the assigned words have been exhausted. If the assignation of line example/pattern is modified, the assignation of words may have to be revised. This new word assignation is now not only constrained by the words assigned to previous fragments, but also by the words already assigned to following fragments. However, if the generation process for any previous line has already finished at that point, any words of its fragment of the intended message that have not been used can be considered.

3.4.3 Revise step

As soon as all the generation processes have concluded, the whole poem draft is presented to the user. The user is required to validate each of the lines. If the user comes up with possible modifications, the system accepts them instead of the generated verse. The modified versions suggested by the user ought to be tested for metric correctness. This feature is not currently contemplated, under the assumption that any corrections the user makes will either be well founded or they will be in repair of a serious syntactic or semantic error introduced by the heuristic approximation.

3.4.4 Retain step

All verses validated by the user are processed to create a personal data file containing the corresponding line examples/patterns in the correct format. Any new words introduced by the user either as part of the intended message or modifications of the proposed draft are also added to the data. This will ensure that regular use of the system will improve the quality of results only if the degree of satisfaction with the initial poems is reasonable high. This implies that although the system is capable of a certain amount of learning once a reasonable standard (of vocabulary and stored examples) has been achieved, an initial threshold of vocabulary and corpus adequacy must be met for the system to function adequately.

3.5 ASPERA Results: Examples

The following examples illustrate system input and resulting poem for specific instances of use of the ASPERA system. Suppose the user expressed a desire for a short, fully rhymed, formal poem, with a rural setting and a positive mood. The following specification was given to guide the search in example 1:

tan hermoso viento fue corazón mudo

The system suggested a *terceto* should be attempted (lines 11 syllables long rhyming ABA), and provided a set of model poems from its case base. Model poems were of the form given in example 0:

Sabed que en mi perfecta edad y armado
con mis ojos abiertos me he rendido
al niño que sabéis ciego y desnudo.

Example 0. Model poem

Since rhyme restrictions are imposed by the generation mechanisms independently of the model poems, these need not follow the rhyming pattern required. The model poems provide guidance as to the number and specific POS elements to be used in constructing each line. Vocabulary files appropriate to the setting and mood where loaded and the system produced the following poem:

Ladrará la verdad el viento airado
en tal corazón por una planta dulce
al arbusto que volais mudo o helado.

Example 1.

Three -*viento, corazón* and *mudo* - of the six words in the specification have made it into the twenty words of the final verse. Six words of the poem originate in the selected vocabulary -*ladrará*, *planta*, and *arbusto* through the rural requirement and *dulce*, *verdad*, and *volais* from the positive mood. Two words - *airado* and *helado* - come from the model poems. These words are forced to appear in the poem by the additional restrictions posed by the rhyme on the final words of the first and last line. As inexperienced human poets do, ASPERA simply chooses two words that rhyme and happens to find more of those among the model poems[1]. The rest of the words form part of ASPERA's elementary vocabulary and are used to hold the poem together with an approximation of sense. In this particular case, the last line of the resulting poem closely follows the structure of the last line of the sample model poem presented. The structure of the other lines is based on different model poems, chosen by the planning stage of the system to ensure minimal overall coherence of the poem.

A similar analysis can be carried out for example 2:

Andando con arbusto fui pesado
vuestras hermosas nubes por mirarme
quien antes en la liebre fue templado.

Example 2.

resulting from similar general choices and the specification:

---

[1] This behaviour may improve if enough rhyming words are provided either in the selected vocabulary or the specification.

andando por el monte hermosas nubes una liebre

The poem as it stands is quite obscure, however, it can be seen very clearly in this case how the specification and the motives - rural and positive - drive the system.

4. Benefits to be Expected from ASPERA

ASPERA may prove beneficial to a certain extent in two different contexts. Within the restricted domain of tuition in the field of literature, and more widely where it may shed light on general language problems.

4.1.1 As a Pedagogical Tool

In the long run, it is hoped that ASPERA will provide guidelines to develop better pedagogical tools for the field of Spanish poetry. For the average person, a poem is either pleasing or not, and the introduction of - sometimes very complex - rules to be computed as part of its enjoyment goes against the grain on first time acquaintance. ASPERA provides the means for students to see how the imposition of the formal rules may affect the form of a message of their choosing, without requiring them to have learnt all possible combinations of the rules. It is expected that interacting with the system may help the students to develop - without learning the formal rules - the instinctive feel for 'correctness' of a verse that a poet achieves naturally.

4.1.2 As a Potential Source for Slim Language Processing Heuristics

The solutions applied in ASPERA to language problems are all highly heuristic in nature, and resort to no complex linguistics techniques. The examples presented show the advantages and the shortcomings of the approach: message content gets scrambled for the sake of form, yet applications relying very heaviliy on specific formats and not requiring originality may find the simplicity of the techniques compensates for the loss of precision. From this point of view the ASPERA system, as it is evolved to more refined heuristics solutions to language generation, constitutes a benchmark for simplified approaches.

5. Conclusion

ASPERA is the computing heart of an application with a great potential for user satisfaction. In its current version, it is handicapped by clumsy interfaces that are unable to compete with state of the art GUI technology. However, the underlying methodologies and techniques have shown their adequacy to the task even in their budding state of development. The heuristics techniques applied to resolve linguistic problems at all levels - syntactic and semantic - open interesting avenues of research in language processing. The system is open to enhancement and adaptation once user feedback has been gathered during evaluation.

References

[1] Aamodt, A. & Plaza, E. (1994). Case-Based Reasoning: Foundational Issues, Methodological Variations, and System Approaches. AI Communications, 7(i), pp.39-59.

[2] Bailey, P., 'A reader-based model of story generation', Symposium on Artificial Intelligence and Creative Language: Stories and Humour, AISB'99 Convention, 6th-9th April 1999, Edinburgh College of Art & Division of Informatics, University of Edinburgh.

[3] Chevreau, K, Coch, J., García Moya, J.A., and Alonso, M., 'Generación multilingüe de boletines meteorológicos', Procesamiento del Lenguaje Natural, SEPLN, N. 25, September 1999, pp 51-58.

[4] Gervás, P., 'WASP: Evaluation of Different Strategies for the Automatic Generation of Spanish Verse', in: Time for AI and Society, Proceedings of the AISB-00 Symposium on Creative & Cultural Aspects and Applications of AI & Cognitive Science, 17th-18th April 2000, University of Birmingham, England, pp 93-100.

[5] Gervás, P., 'Un modelo computacional para la generación automática de poesía formal en castellano', in: Actas del XVI Congreso de la SEPLN (Sociedad Española para el Procesamiento del Lenguaje Natural), Vigo, España, September 2000.

[6] Gervás, P., 'A Logic Programming Application for the Analysis of Spanish Verse', in: Lloyd, J., et al, Computational Logic - CL 2000, First International Conference, London, UK, July 2000, Proceedings, Springer Verlag.

[7] Horacek, H. and Busemann, S., 'Towards a Methodology for Developing Application-Oriented Report Generation', in: Günter, A. and Herzog, O. (eds.), 22nd German Conference on Artificial Intelligence (KI-98), Proceedings, Bremen, Germany, 1998.

[8] Hisar Maruli Manurung, Graeme Ritchie and Henry Thompson, 'Towards a computational model of poetry generation', in: Time for AI and Society, Proceedings of the AISB-00 Symposium on Creative & Cultural Aspects and Applications of AI & Cognitive Science, 17th-18th April 2000, University of Birmingham, England.

[9] Nederhof, M.-J., 'Efficient generation of random sentences', Encyclopaedia of Computer Science and Technology, Vol.41, Marcel Dekker, 1999, pp 45-65.

[10] Quilis, A., (1985) 'Métrica española', Ariel, Barcelona.

[11] Real Academia Española (Comisión de Gramática), (1986) 'Esbozo de una nueva gramática de la lengua española', Espasa-Calpe, Madrid.

[12] Sánchez León, F., Corpus Resources And Terminology ExtRaction project (MLAP-93/20), 'Spanish tagset of the CRATER project', ftp://ftp.lllf.uam.es/pub/CRATER/esT-tagger-1-0.tar.gz

SESSION 1

KNOWLEDGE MANAGEMENT AND KNOWLEDGE ENGINEERING

Creating Knowledge Structure Maps to support Explicit Knowledge Management

John L. Gordon: Applied Knowledge Research Institute, Blackburn College, Feilden Street, Blackburn, England. Email john@akri.org

Abstract:

The idea that knowledge management can be carried out explicitly is often challenged by many in the knowledge management discipline. Some even challenge the value of the term 'knowledge' in this domain and have instead widened the definition of knowledge management to include almost anything that improves the business. Although it is accepted that many of the methods and the systems described within the knowledge management domain can be of great value to organisations and are clearly sensible steps to take, it is not necessary for those approaches to be exclusive of the possibility of a discrete approach to the direct management of the knowledge resource.

One of the first questions asked by those taking their first steps in knowledge management is 'What is knowledge?'. It is not necessary to avoid this question and a look at what the answer may be can also show why it may prove very useful to ask the question.

The approach of Structural Knowledge Auditing and Mapping the Structure of Knowledge is intended to provide managers with a visualisation that can become a common frame of reference against which to discuss the knowledge resource. Including several (rather subjective) parameters that describe each of the knowledge nodes on the map can also form a basis of a very useful way to analyse the knowledge resource and focus decision making on to areas of greatest need. The method can provide a way to create demand for other knowledge management strategies including Knowledge Based Systems, GroupWare, Knowledge Warehousing etc.

This method has already been applied successfully in several large companies in the UK and also some smaller companies and even in the off-licence retail trade area. In most cases the work has also lead to the management of the organisations in question making significant changes that have been identified from the results of the audits.

Acknowledgements:

- This work has been partly funded through an ESF ADAPT project.
- Thanks to Blackburn Regeneration Partnership for making necessary resources available to complete this work.
- Thanks to Brian Whitehead – Dean of Creative Arts – Blackburn College – for help with the final draft.

1 Introduction:

Much of the published work on knowledge management actually questions whether the term can really mean the management of knowledge and much of it redefines the term to cover a very general range of business activity. The general activities including Data Warehousing, GroupWare and Intranet solutions have been shown to provide real business benefit. However, some authors promote these systems by suggesting that any explicit management of knowledge by managers is not possible anyway.

1.1 What has been said about KM?

The following collection of statements made about knowledge management has not been attributed to any particular publication or author to avoid causing embarrassment. However, the sources are available.

"... knowledge management is a misnomer from the start because its not about knowledge capture but about information capture to improve knowledge."

"... knowledge management is an oxymoron. Knowledge can't be managed and once you switch to this higher level of 'knowledge' all the situatedness gets thrown away."

Admittedly the two statements above are out of context but they do show that these authors do not believe in the possibility of explicit knowledge management. It is not clear whether they are not sure what knowledge actually is or if they do know this, why they have concluded that it cannot be managed.

"Knowledge Management, as a term, is only important as a means of grasping the latest stage in organisational theory and corporate development."

".... illustrate a fundamental concept of KM: the use and re use of data in an organisation to feed the whole so called 'business process'."

The statements above show that these authors consider knowledge management to be an indefinable concept that is really used to describe a 'successful' business strategy.

"Good knowledge management is getting the right information to the right people at the right time."

"Knowledge management is a model for laying forth the groundwork needed to maximise returns from intellectual capital."

These statements are more clear about the use of knowledge to achieve certain objectives but still do not tackle knowledge management in any more focused way.

"One of the huge limitations of knowledge management has been the pitfalls of full text searching."

This statement again implies that knowledge management relates to anything that can be documented and avoids a direct approach to the management of knowledge.

"KM + Internet = ebusiness [2]"

I have no comment to make about this statement.

A quotation from 'Murphy's Laws' (Bloch 1985) seems to be appropriate here:

"The one who says it cannot be done
should never interrupt the one who is doing it."

1.2 Knowledge Management as a business strategy

The statements in 1.1 show that knowledge management is growing as a business strategy. This is not a bad thing because it does focus effort on a valuable organisation resource. This approach is also supported by some excellent software that can be used to achieve fairly rapid results and can give a focus to the general company strategy of knowledge management.

The problem is that many of the statements above are highlighting the benefits of a particular approach, in part by steering companies away from any direct and explicit approach to knowledge management. There is no need to do this. This paper will show that one direct and explicit approach to knowledge management can lead managers to design and implement these more global strategies. More importantly, the managers concerned will know why they need these approaches and will have already shown what they are intended to achieve.

Helping managers to take a more direct control of the knowledge resource is likely to lead to more general knowledge management activity and not less. The knowledge resource should feature in the decision making process at all levels. It is often the more strategic decisions that are unable to draw on any useful information about the knowledge resource. For instance, when the board of directors discuss the outsourcing of many of the business IT functions, it should be possible to get a direct answer to the question 'How will this affect the companies knowledge resource?'. At other levels, when the section manager is asked to show a team from a competitor company round the manufacturing process, it should be possible for that manager to ask in advance, 'Exactly what part of our knowledge resource will we be sharing with these visitors?'.

1.3 A method of direct control

The Applied Knowledge Research Institute has been developing a methodology for mapping the structure of knowledge and using the resulting visualisation and data to offer a way of supporting decisions involving the knowledge resource with verifiable analytical data. This work has developed from earlier work started by the North West Artificial Intelligence Applications Group in 1994. This earlier work considered the totality of a knowledge management scheme and drew on a broad range of expertise from over 14 organisations (Gordon, Edge 1997). One of the major aspects of this work was to consider what a knowledge management scheme should be expected to deliver, given that the work started in 1994 and was not called knowledge management then.

The earlier project lead to some excitement about the possibilities for knowledge management and certainly its potential was appreciated. Work in what could be described as the main stream of knowledge management is still on going and a collaborative group continue to develop this aspect (Gordon, Smith et al. 1999). A knowledge management booklet for business has also been produced to try to clarify this area of work (Gordon, Smith 1998). Difficulties of direct management control of the knowledge resource and also the lack of rigour and justification in some aspects of the then growing area of knowledge management were also identified. Such difficulties caused us to take a more detailed look at the whole area of knowledge and at what direct management of a knowledge resource could mean.

The main thing that makes knowledge difficult to manage directly is a lack of some frame of reference or an adequate representation scheme. Coupled with a general lack of understanding of what knowledge actually is, it is not surprising knowledge management has no real focus. However, it is also true that managers are now (in general) better trained and better equipped to manage than they ever were in the past. The obvious thing would be to help these managers understand, visualise and take control of the knowledge resource in a very direct way. The work described in this paper aims to do that and attempts to achieve these direct results by focusing on a map of the structure of a knowledge area that employs learning dependency as its structural integrity (Gordon 1999).

2. Knowledge

Before tackling the relatively easier knowledge management related issues, it is interesting to first explore the rather contentious question of, 'what is knowledge'. In exploring this it would also be useful to ask the question, 'Why do I need to know what knowledge is?'.

The first question often asked by people new to knowledge management is, 'What is knowledge?' and unfortunately it is the question that is either least well answered or not answered at all. Usual attempts to answer the question end up in some sort of political debate style opt out as the question is redefined or put aside for a more important one. Needless to say, any real attempt at answering the question is bound to be controversial.

2.1 What is knowledge?

It is not easy to find a simple definition for knowledge. Dictionaries in general do not provide one that is useful. The Collins Dictionary of Artificial Intelligence states that knowledge includes facts, beliefs and heuristic rules. More standard definitions for knowledge found in Artificial Intelligence texts are similar to the one found in (Graham & Jones 1988). This explains knowledge by pointing out that data can be a number (1000), information provides context to data (1000 isobars at noon) whilst knowledge is more (Most people feel better when the pressure rises above 1000). Philosophy is a richer source of definition for knowledge and even separates this out into the discipline of Epistemology. The book, Epistemology, an Anthology, is a comprehensive source of information about knowledge for those wishing to attain an in depth understanding. The work also contains simpler definitions (Sosa 2000). Traditional definitions state that knowledge is:

'Justified True Belief'

Things are not quite that simple however because it can be shown quite clearly that there are occasions that someone can be justified in believing in something that is also true and yet this is not knowledge. There is not space here to go into this sort of analysis.

These definitions remain inconclusive and reasoning can become circular when one looks at the very basis of what can really be believed. However, for the purposes of working with knowledge, there are some useful observations that can be made.

1) If something is not true then it is not knowledge
2) There needs to be some justification for believing that something is true
3) Knowledge does not have to be complex although much of it is

2.2 Is it important to know what knowledge is?

The three statements identified in section 2.1 have implications for knowledge management. For instance, if a company knowledge archive is created and there is no way of preventing things that may not be true getting into the archive then statement 1 is not satisfied. Even worse, at some time in the future the knowledge archive may become worthless, containing as much that is not true as things that are true.

A knowledge archive that provides no justification for pieces of knowledge may also be of little use to the human users of the archive. A justification may set important constraints on the truth of the knowledge that would prevent it from being used in an inappropriate context.

The conclusion to be drawn here is that although many authors are convinced that knowledge itself is an unimportant distraction in knowledge management, a complete ignorance of knowledge may be responsible for the design of knowledge management strategies whose usefulness will be relatively short lived.

Some definitions of knowledge show that it represents an increase in complexity from data, through information to knowledge. Statement 3 in section 2.1 suggests that this is not the case.

In short, it is still not easy to clearly define knowledge but all knowledge workers should be aware of the three statements shown in 2.1.

3 Creating and Using Structural Knowledge Maps

The idea of a map that shows the structure of knowledge rather than the knowledge itself is something that has been defined here. The main point is to provide a visualisation for a knowledge domain. In addition, the visualisation will have an independent value because of the method used to create the structure. The knowledge can be managed directly using the map as a focus for discussion and there are fewer implications concerned with definitions.

3.1 Concept Maps

Concept diagrams / maps are closely related to semantic networks. Concept diagrams are also composed of nodes and arcs that have similar functions. Concept diagrams can be used to describe fairly complex concepts and are suitable for both machine and human interpretation. They are seen as a knowledge representational method that employs graphical structures (Sowa 1984). There is a body of work relating to concept diagrams and their use as a graphical logic (Sowa 1993). This offers interesting opportunities for future work on knowledge mapping by creating the framework that could allow knowledge maps to be transformed into other machine understandable representations such as the Knowledge Interchange Format (KIF) (Genesereth 1992).

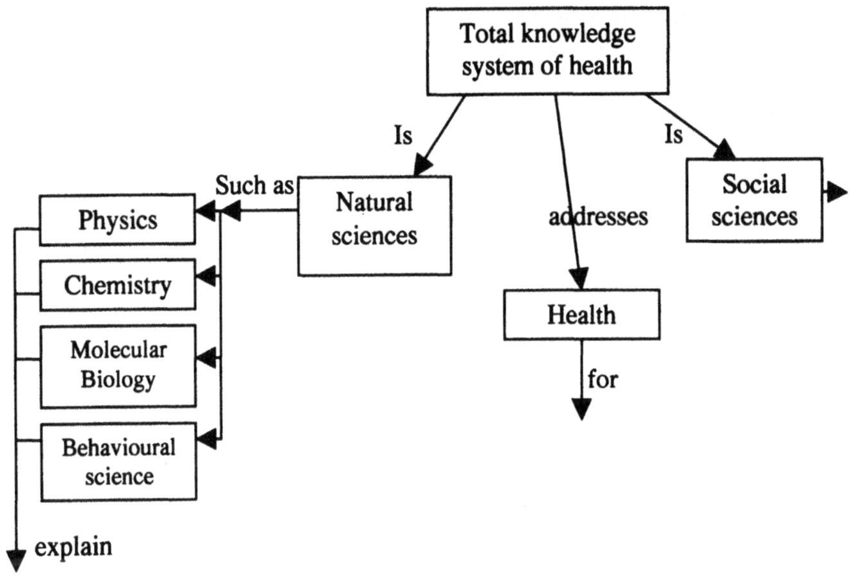

Figure 1: Extract of Concept Diagram for knowledge system of health University of Florida – College of Medicine.

Concept diagrams are a powerful way of representing knowledge, of creating a common visualisation and also a powerful learning tool. Figure 1 shows a very small portion of a diagram created by the College of Medicine at the University of Florida. The diagram does not require too much explanation because explanatory sentences can easily be created from the diagram itself. The main point to note as far as this reference is concerned is that each arc (or line) has a label that specifies the function of the line. In the work to be described here, arcs have one common function.

3.2 Learning Dependency

The common function of an arc for our type of knowledge structure map is 'Learning Dependency'. Each arc on the knowledge structure map is directional and shows that the target item of knowledge must already be known before it is possible to fully understand (or possibly learn) the source item of knowledge.

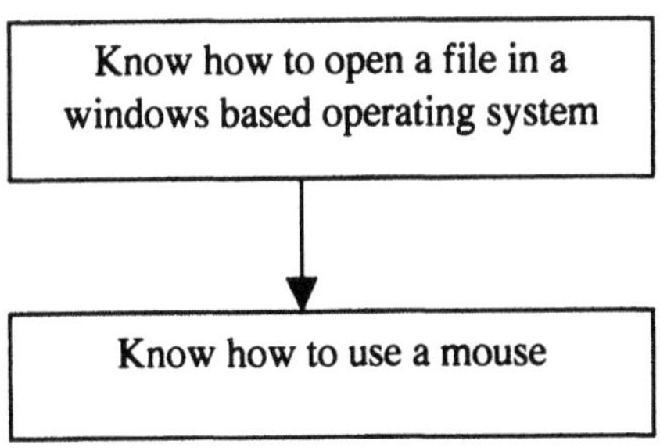

Figure 2: Illustration of learning dependency

Figure 2 shows that in order to know how to open a file in a windows based operating system it is first necessary to know how to use a mouse. It could of course be argued that this is not the case and it is only necessary to know how to use a mouse in order to actually open a file not to know how to do it. However, the concepts of mouse movement and related pointer movement and button clicking would be meaningless without the prior knowledge of how to use a mouse.

Learning dependency is a human centred approach to mapping the structure of knowledge. Learning dependency means that it is necessary to know knowledge 'Y' before knowledge 'X' can be fully known.

Learning dependency also provides at least some encouragement that it is really knowledge that is being managed. The dependency structure, as used when acquiring expert knowledge, provides some justification as to truth value. One of the reasons I am justified in believing 'X' is that I already have knowledge 'Y' and 'Z' and I know that the knowledge of 'X' is dependent on a knowledge of 'Y' and 'Z'. This does not fully satisfy the first two statements discussed in section 2.1 but it does go part way towards this.

3.3 Attaching parameters to Knowledge Nodes

Interviews are used as the main vehicle in the construction of structural knowledge maps. During interviews, people are asked to provide information about the structure of knowledge in the domain in question. There are many possibilities as to what this information should be and of course an acknowledgement that interviewees will often need to estimate answers based on their own experience. Several parameters and combinations of parameters have been tested and the following four have been found to be the most useful in all audits.

Importance	How important is the knowledge to the company?
Difficulty	How difficult would it be to replace this knowledge?
Study – Experience	Is the knowledge acquired mainly from study or practice?
Known By	What proportion of the staff in the knowledge area know this?

Each of these parameters will have a value attached between 0 and 9.

0 -> unimportant, not difficult, non of it etc.

9 -> core knowledge, very difficult, all of it etc.

Parameter values are estimates and can be subjective. However some validation does occur during the interview process and it is important to inform managers that the parameters reflect what their staff think and if this is a problem then this may also be something that requires attention.

3.4 A map of the structure of knowledge

The interview procedure is fairly rigorous and consists of group and individual interviews. The outcome of the process is a structural knowledge map based on learning dependency with parameter values attached, as identified in section 3.3. Maps are either viewed with a computer based tool or can be printed out, often on A0 size paper to be hung on a wall. A typical map will consist of between 60 and 100 knowledge nodes. In order to investigate a knowledge area in more detail than this it would be better to work on separate, linked maps rather than have maps of more than 100 knowledge nodes.

It is difficult to reproduce a full map in a paper like this but the idea can be conveyed through samples. The next two figures are from a self audit of the AKRI which was carried out for demonstration purposes but which actually provided useful information. Figure 3 shows a part of the map without parameter figures. Figure 4 shows the full map but no detail is included.

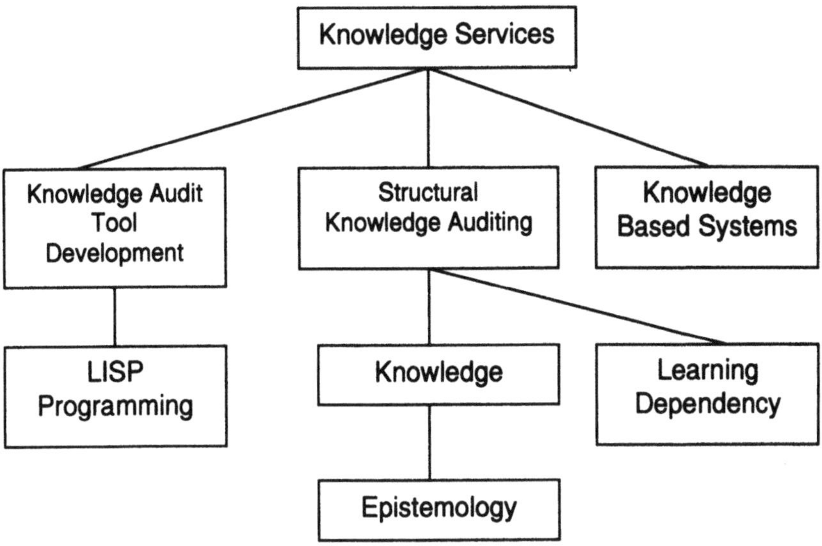

Figure 3: Extract from a knowledge structure map

The extract shown in figure 3 is for illustration only and several peripheral arcs and nodes have been removed for clarity and the layout has been changed from that shown in figure 4.

The figure shows that in order to have a full knowledge of the AKRI knowledge services, it is necessary to first have a knowledge of 'Knowledge Based Systems', 'Structural Knowledge Auditing' etc. In order to know how to do 'Structural Knowledge Auditing' it is necessary to know about 'Learning Dependency' (and other things of course).

The complete strategic map of the structure of the AKRI knowledge is shown in figure 4.

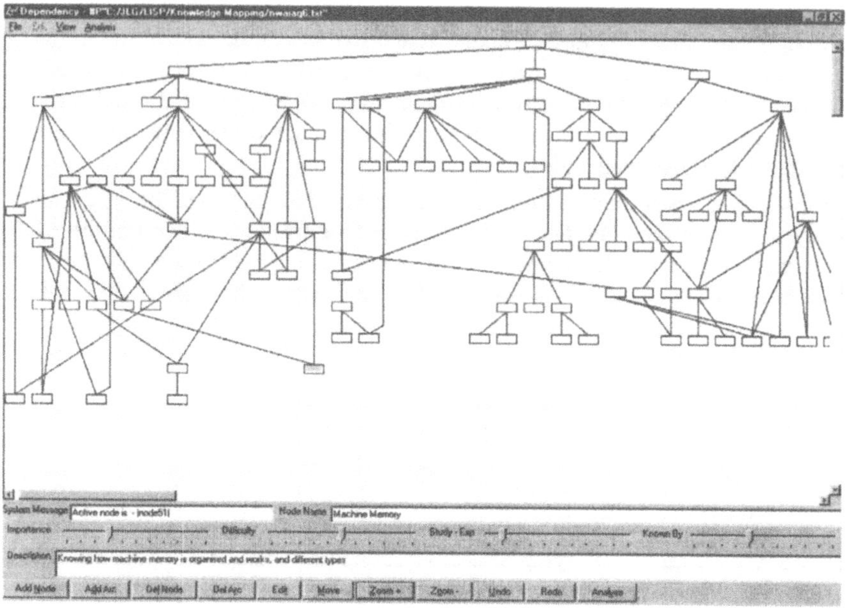

Figure 4: Strategic level map of the structure of AKRI knowledge

Detail has been omitted from the map but the layout of the support tool has been included to provide an idea of how parameter elicitation is coupled with the elicitation of the structure of knowledge (the software is being written with Harlequin LISP).

A structural knowledge map derived in this way does not really have to satisfy statements 1 and 2 in section 2.1 because the map only contains the structure and not the knowledge itself. Managers can manage the knowledge resource by using the structure and leave the truth value to the experts in the knowledge domain.

3.5 Analysing the results for management decision making

The software tool that is being developed to support the interview and map creation process, also supports a capability to provide statistical analysis of the map, its structure and the parameters assigned. The software provides tables of data that are sorted in various ways.

For instance:

- A list of all nodes in alphabetical order along with their parameter values
- A list of all nodes and parameters in 'importance' order
- A list of all nodes and parameters in 'difficulty' order
- A list of all nodes and parameters in 'study-experience' order
- A list of all nodes and parameters in 'knownby' order

- A list of all nodes and parameters in 'risk' order
- A list of nodes in alphabetical order along with connectivity data
- A list of nodes with connectivity data in highest prerequisite connectivity
- A list of nodes with connectivity data in highest postrequisite connectivity
- A list of all node names, prerequisites and associated notes

A table showing the most important knowledge or the knowledge known by the least number of staff is useful to support management decision making.

In this case, risk is a derived value and provides additional information for managers to help guide them towards knowledge that may require their urgent attention. In the current system, risk is computed as follows.

$$\textbf{Risk} = \frac{I + D + S + (10 - K)}{4}$$

Where: I = Importance
D = Difficulty
S = Study / Experience
K = Known By
Values between 0 ... 10

Within the context of this work, a risky piece of knowledge is one that is very important to the organisation, very difficult to replace, mainly experience based and known by only a few staff.

There is not space here to fully describe all of the information that the process produces. However, it is important to state that all of the information is intended to inform managers about the state of the knowledge resource and allow them to use their management skills to take decisions and plan action based on the analysis.

4 Structural Knowledge Auditing in Business

This methodology has been used in three large companies, one small company and two other business organisations. During each of these projects, the methods have been developed and refined. However, in each case there have also been specific benefits realised and further work is either in progress or being planned involving new companies (large and small) and a plan to return to previous organisations for additional work.

4.1 The earlier projects and methodology refinement

The earliest full scale structural knowledge audit was in a large manufacturing company and concerned a particular specialist manufacturing section. The company managers had some concerns that they were part of a large project consortium that included a knowledge sharing agreement. The problem was that they did not know what knowledge was being shared or even what knowledge was part of the specialist manufacturing area in question. The knowledge map and subsequent analysis caused a great deal of debate amongst the management group and as a result, the knowledge sharing and visiting arrangement were more precisely defined and controlled.

The main areas of methodology change that resulted from this audit was a tighter definition of the requirements of the knowledge audit support tool and the realisation that, for the present time at least, a commercial system would not do the job. The interview and initial presentation parts of the process were also more clearly defined.

4.2 Working in larger companies

Two more audits were carried out in large companies, one in a large utility company and the other in a multinational engine manufacturer. The audit in the utility company highlighted several important problems with the techniques used. The main one of these was the problem of separating expert level understanding from a knowledge of answers to expert level questions without any background understanding. This in turn lead to a greater research effort in the area of knowledge and the realisation that knowledge must be understood well by the interviewing and analysis team even if it is not well understood by the staff in the company. Tighter controls of the audit methodology solved problems that arose during the utility company interview.

The interview in the third large company used custom design support software and employed a tighter methodology. This better control certainly proved effective and the audit was a considerable success. One of the most rewarding parts of the process came when presenting results. Over a break period, several managers stood around the structural knowledge map, pointing at areas and discussing them with a common frame of reference. This discussion continued around the table and it was difficult to stop the managers discussing their knowledge resource so that we could get continue presenting the results. It was accepted by the management team that this was the first time they had ever discussed anything that could be shown to be the knowledge resource directly and explicitly.

Even though this audit was very successful, it was still possible to identify improvements and implement them. The main area of improvement concerned the analysis of the map data. The analysis procedure was more precisely defined and it was possible to specify this analysis well enough to implement software support that automatically produced map analysis data (section 3.5). This replaced a great deal of the manual effort that went into creating a report and also supplied more rigour and control to the analysis process itself.

4.3 The needs of smaller companies

Contact with smaller companies of around 60 to 200 employees has revealed an interest in knowledge management but there is a difficulty in defining a practical implementation strategy. It is often the case that the smaller company cannot justify any significant investment until a clear payback pattern is identified. This does not mean however, that the smaller company is ignorant about the value of its knowledge resource.

The Structural Knowledge Audit technique can provide an incremental approach to managing the knowledge resource with specific deliverable phases that can also be used to guide future phases of the work. There is an on going project involving a small manufacturing company with about 100 employees. This involves carrying out a specific audit in the key engineering area. The results of this will be available within 3 weeks of the audit and will be used to support decisions about the next audit phase. The current work will involve three incremental phases and cover areas selected from the whole business. The management team in the company will not only benefit from the results of the audit but will be able to use results from early work to guide later audits to areas of greatest need.

4.4 Other business groups

In one project, an audit was undertaken at a strategic level in the off-licence retail trade area in the north of England. It was expected that this work would be particularly demanding because the people interviewed were from separate businesses and had not met each other before the initial presentation. It was expected that the knowledge used in the separate businesses could be difficult to unify in a common structural knowledge map. This did not prove to be the case however. The process of presentation, group elicitation and individual interview programme with validation support proved to be up to the demands of the diverse knowledge area. It was also clearly a great advantage that the people interviewed became very interested in the idea of having the knowledge that was used in their area of business mapped out and open to discussion. Many were surprised that the knowledge area was so extensive.

5. Summary of the approach

The aim of the structural knowledge auditing approach is to produce a map of the structure of knowledge in a specific area. This map can show users any level of detail depending on the intended uses of the map. For instance, the board of directors may want answers to strategic questions about the knowledge resource of the whole company but are unlikely to want to know about the knowledge at a detailed operational level. The training manager may want to use a map to plan training and to help organise training in areas that have been shown to be at greatest risk to the company. The manufacturing manager may want a clear visualisation of the knowledge used in a particular manufacturing area and may want detailed analysis so that actions can be planned to protect and develop the knowledge asset.

The method can make information about the knowledge resource available to managers at many levels. Decisions based on this information will also be part of the knowledge management process. It is possible that managers looking at a strategic level may be concerned that some important areas of knowledge are not widely available and may decide to implement a GroupWare solution across the company. Managers of a particular manufacturing area may be concerned about the high risk of a critical piece of knowledge and investigate the development of a Knowledge Based System to provide consistent reliable knowledge anywhere that it is needed.

The point of these scenarios is that managers themselves will be the ones who are able to discuss the knowledge resource based on an analytical visualisation and decide on the correct course of action to take.

It is argued in this paper that it is possible to discuss knowledge, to know what knowledge is and to manage it directly and explicitly. Structural Knowledge Auditing (as described in this paper) may not be the only way of achieving this but it is certainly one way of doing so.

References

Block A. Murphy's Law, Complete. Mandarin. 1988, pp211

Genesereth M.R. Fikes R.E. Knowledge Interchange Format, Version 3 Reference Manual. 1992. Report Logic-92-1, Computer Science Department, Stanford University.

Gordon J.L. Edge M. Focused Knowledge Management, in Applications and Innovations in Expert Systems. Ed. Macintosh A. Milne R. SGES Publications 1997 pp207,219

Gordon J.L. Smith C. Knowledge Management Guidelines. NWAIAG Publication. 1998. (available from the AKRI)

Gordon J.L. Smith C. Scott S. McBriar I, Walbank S. Bain G. Unsworth P. Practical Approaches to Knowledge Management in Applications and Innovations in Intelligent Systems VII, Eds Ellis, Moulton & Coenen, Springer, 1999 pp 17-32

Gordon J.L. Creating Knowledge Maps by Exploiting Dependent Relationships in Applications and Innovations in Intelligent Systems VII, Eds Ellis, Moulton & Coenen, Springer, 1999 pp 63-78

Graham I, Llewelyn Jones P. Expert Systems Knowledge Uncertainty and Decision. Chapman and Hall 1988 pp 20-23

Sosa E, Jaegwon K. Epistemology An Anthology. Blackwell pp 55-57

Sowa J.F. 1984. Conceptual Structures: Information Processing in Mind and Machine. Addison Wesley 2000

Sowa J.F. Relating Diagrams to Logic. in Concept Graphs for Knowledge Representation Eds Mineau G.W., Moulin B. Sowa J.F. Springer-Verlag 1993. pp1-35

Towards Continuous Knowledge Engineering

Klaas Schilstra and Pieter Spronck
Knowledge Based Systems Department
TNO, Netherlands Organisation for Applied Scientific Research
Delft, The Netherlands
{k.schilstra; p.spronck}@bouw.tno.nl
www.tno.nl

Abstract: Continuous Knowledge Engineering is an alternative approach to knowledge engineering that embraces the philosophy that knowledge systems are open-ended, dynamic artefacts that develop through a learning process in reaction to their environment. This approach is based on the debate on symbolic and situated models of human cognition and aims to make knowledge engineering a learning process by adherence to incremental development, participation of the expert and direct application of the knowledge system. This paper presents this approach to knowledge engineering, examines a knowledge modelling tool that supports it and looks at an application of this tool.

1. Introduction

There has been extensive discussion on the symbolic or *descriptive* stance versus the situated theory of human cognition [1,2]. This discussion has opened up a re-evaluation of the physical symbol system hypothesis that has been held as defining for knowledge system development and knowledge engineering from its very conception. The hypothesis defines a theory of cognition based on explicit representation of knowledge and the goal-oriented or rational use of this knowledge by agents [3].

The situated cognition hypothesis instead proposes that all knowledge is grounded and connected to the context in which it is acquired and applied. This would imply that as knowledge is used it will change with each new experience and that static representations of knowledge degrade. This questions many of the assumptions that underlie knowledge engineering today. The proponents of *strong situated cognition* profess that only reactive systems, based on learning and inductive approaches, devoid of representation and reasoning would therefore have sufficient and necessary facilities for intelligent action [4, 5].

A middle way is *weak situated cognition* [1], which says that as knowledge changes through experience, an approach that permits a knowledge model to follow that change may circumvent the adverse effects. In addition, this would allow knowledge engineers (KE) to use current approaches but with a change in perspective on the employment of those techniques. In short, knowledge engineering must be a learning process, bound to the practical use of the system. This would seem to be a natural progression from the current model-based approaches towards an approach that embraces change in a knowledge model through learning.

This paper will first give a short overview of currently experienced problems and aspires to provide some explanations for these problems. In addition a knowledge engineering approach based on a learning process is proposed that aims to circumvent or solve these problems. A tool is described that supports this approach to knowledge engineering and an application of the tool is discussed. Finally some conclusions are reached and some future work is discussed.

2. Problems in Knowledge Engineering

The state-of-the-art in knowledge engineering is to a great degree a result of the symbolic view of knowledge and cognition. It offers a constructive model of human reasoning as operations on symbol patterns. In this way it has been a successful cognitive model leading to useful systems. In concordance with the statements by Clancey [2], these accomplishments should not be detracted from. From the earliest systems to the systems developed using mature methodologies, such as CommonKADS [6], certain problems are associated with the development of knowledge systems.

From a practical and commercial perspective the most important problem is the cost and risk associated with knowledge system development. This makes it difficult to 'sell' knowledge systems as viable solutions to problems that organisations may experience, and creates a situation where knowledge systems are still perceived as experimental technology. Not many companies allow mission critical applications to be based on knowledge technology as yet, although there are some notable exceptions, for example XCON, which supports DEC's configuration management and has done so for many years [7]. This system has further displayed considerable benefits, saving the company millions a year in economic terms. But despite these successes and many others, war stories from the industry often relate of projects that do not rise above initial prototypes or unsuccessful introductions of finalized systems.

The cost and risk can be attributed mainly to the intrinsic difficulties of developing knowledge systems. The two main culprits are the knowledge acquisition and maintenance problem and the gap between prototype systems and industrial strength knowledge systems [8].

The fact that knowledge acquisition is hard is well known. The knowledge acquisition bottleneck is caused by the inherent difficulties in formulating and formalising knowledge, intensified by communication problems between the KE and the expert [9]. The expert and KE have to 'learn' to work with each other and create a combined view of the domain, to be able to structure the domain knowledge, identify important concepts, relationships, rules and heuristics.

After the development of a knowledge model and system, maintenance of knowledge models to remove flaws and to complete missing features is performed.

This is not commonly perceived as the natural degrading of knowledge or the shift in meaning caused by its situated nature. However, knowledge systems show considerable change over their lifetime. XCON shows that 50% of the rules may be modified, deleted or added during a given year [7], which can be explained as resulting from new products or configurations. As was shown more recently in the PIERS system, knowledge changes considerably even when the domain itself is static [10, 11] (see also below). The maintenance of knowledge system therefore contains a clear and sizeable component of knowledge acquisition and modelling.

The gap between a successful knowledge model and the development of a user-friendly, industrial strength knowledge system constitutes a further problem. Even after a successful knowledge model has been developed, the ability to translate this model into an application acceptable to the end user is tenuous. Besides misconceptions and neo-Luddite fears, the users often find the level of such applications to be below par. The effort to develop user-interfaces of a quality that most users have grown accustomed to and inclusion of auxiliary facilities, such as integration with existing information systems, can require even more work than the total effort of developing the knowledge model itself. This is discussed in literature [8] but also has been observed in applications developed by the authors. The percentages seem to show that this additional development constitutes from 50% to 80% of the total development effort.

Current research approaches aim to address the knowledge acquisition and maintenance problems mentioned above by increasing the control and management of knowledge engineering, usage of deep models and reuse of existing knowledge models. Problems with the development of actual applications are seen as a tooling problem. A prime example of this is CommonKADS [6], which places much importance on methodology and constructing the right model, and implementing it. The model library of CommonKADS is extensive and allows for reuse of both problem solving methods (PSMs) and domain models. Knowledge models are developed using a risk-driven spiral development process. The goal is to develop a model that goes towards a final system that can be maintained for flaws and incompleteness in its functionality. CommonKADS further attempts to moderate the problems in development of the final system by giving due attention to the organisation, the technical environment and modelling the communication with the user.

Knowledge system development embracing situated knowledge acquisition is also performed but is much smaller in scale and has yet to be embraced by the mainstream. The results from the PIERS system are exemplary for the benefits that can be gained by using incremental approaches based on grounding the system in its environment [10, 11]. It is also one of the few examples where experts or domain specialists are actively involved in the knowledge modelling process using ripple-down rules (RDR). The system has been developed without any design or model, by the active involvement of the experts and an incremental prototyping approach. The maintenance proceeded on a case to case basis. By spending on average a couple of minutes a day the system was maintained. Every change is immediately part of the operational knowledge system. Using this approach the system operated with considerable success, reaching accuracy levels over 95%.

Considering knowledge acquisition as a process of learning, can explain the inherent difficulty of knowledge acquisition as well as the inordinate amount of maintenance required. If knowledge originates in experience and is intimately tied to the environment in which it is acquired and employed, then this would question the current research approaches. It would place doubt with design-based approaches that would try to create a description of the knowledge and the domain needed to perform within that domain without grounding them in experience. Another of the consequences of accepting the weak situated cognition position is that we can not expect to reuse existing models or PSMs *as is* to bolster the productivity or share the cost of development of a knowledge base. The problems that are experienced in creating knowledge models and the results from the PIERS system seem to indicate that there is at least a place for improvement. If the situated view is taken the focus should be on how we build and change models, to allow room for continuous evolution and exploration, characteristic of any learning process.

Coming back to the earlier comments on the ability to sell knowledge technology as a viable solution, for the art of knowledge engineering similar remarks can be made. If research and development in knowledge engineering can improve the applicability of knowledge systems, and increase the number of experiences in the form of successful and unsuccessful knowledge systems, the volume of these experiences will allow the improvement of the best knowledge of knowledge engineering, as well as further the cognitive theories these systems embody.

3. Continuous Knowledge Engineering

The Continuous Knowledge Engineering (CKE) approach embraces the philosophy that knowledge systems are dynamic, open-ended artefacts that must develop in response to their environment. Therefore, developing knowledge systems requires that a development methodology is adopted that is based on a learning process. Learning is seen as a continuous process of incremental insight, change and application leading to further insights.

The CKE approach is not so much a methodology itself but a change of perspective on the deployment of current methodologies, tools and techniques; a set of ideas that allows a learning process. The approach centres on three principles to accomplish this: incremental development, active participation from non-specialists such as experts, and direct application of the knowledge system. Each of these principles will be discussed in the sections below.

3.1 Incremental Development

Incremental development means developing a following step based on the previous one. Basically, lessons learned in the current iteration can be applied to better enable and inform the tasks in next ones. This is the reason why many evolutionary or incremental software engineering methodologies favour it to enable risk- or functionality driven approaches [12, 13]. As was discussed in the previous there is already a tendency to employ iterative strategies in other approaches to knowledge engineering. But in these approaches the iteration mainly takes place in the design phases, and not through cycles of implementation, application and design.

In many situations a solution to a problem can not be created, but must be revealed through experimentation and exploration. Only when the knowledge is made explicit can certain epiphanies take place, clarifying and illuminating old problems in a new light. And only by positioning the knowledge models in the real world, where they are confronted with real problems and real solutions can we be assured of the quality of that model and gain experiences leading to novel insights. Anything less than this may provide a candidate with which we can delude ourselves. This harks back to the grounded program defined by Brooks [14].

The ability to incrementally develop and ripen a knowledge model, to incorporate new knowledge and improve on the knowledge already present in the system is the pivotal element in the approach. Knowledge builds on knowledge, and true insights can only be gained working and testing our incomplete models of the world and discovering imperfections in its understandings. The focus of the CKE approach is therefore on fast and frequent delivery of the knowledge system. In this way the CKE approach becomes based on the experiences with the product, rather than on detached requirements or models. This is more flexible as the system does not have to constitute a complete solution, as long as it demonstrate evolution and can be used a basis for evaluation of the progress and direction of the development.

3.2 Active Participation

The added insight gained by being involved building an explicit knowledge model is placed with the wrong person, when it is performed by the KE. The person best able to appreciate the possible insights to be gained from the development and application of the knowledge model is the expert. The practical experience of the expert should give the expert a better basis to learn and make new discoveries. The expert is grounded in the field and can bring his or her own practical experiences to bear. The expert will also gain the most benefit from interacting with the model. Making their understanding of the task and the domain explicit will aid experts in deepening their own knowledge.

The KE should focus on the structuring of the knowledge system and facilitating the development of the system. Furthermore, the iterative way of working would require the KE to be either constantly connected to this single project. This could be done if the effort was limited but regular or when the knowledge model would be large enough to warrant such singular attention. But this would not be very effective or efficient and would most likely also not be appealing to such persons.

Active participation focuses on getting the domain specialist to be actively involved in modelling knowledge. This can be in unison with a KE, but the ultimate intent is to get the expert working directly on the knowledge model. The wish to have the expert participate in the knowledge engineering has existed for long as it represents certain financial and practical realities. However, knowledge modelling by the expert is not a common approach in AI, although examples are known (e.g. RDR [10], Spark-Burn-Fire Fighter [11]). From these examples it is clear that it is possible, but there is a considerable need for support from the tools used to create these systems.

3.3 Direct Application

Direct application means that the knowledge model can be deployed to its intended users early in the development and after that with each increment. Simply said, this means that a system based on a single rule should be both consultable and deployable with ease, and that changes to the knowledge model should also be published in rapid succession. Furthermore, the quality of the systems should be such that it is acceptable to its users.

One of the main priorities in development of knowledge systems is that someone will use the system – it has, at least, to meet with interface standards that end-users are expecting from traditional application programs [8]. The fitness of the knowledge system to the expectations of the users is an essential criterion for the acceptance of the knowledge system and the knowledge it contains. This kind of focus on 'business purpose' can also be seen in other incremental or evolutionary development methodologies such as DSDM [13]. There is no reward for delivering the best product, just for the delivering a good solution.

As was mentioned considerable effort goes into developing the visualisation and auxiliary facilities for a knowledge systems as well as integrating the system with existing information systems or as a component of a larger system. Making this easier or spreading this effort over a longer period, may help easing this and discover the best form in interaction with the users and the organisation.

3.4 Conclusions

On the surface, CKE may seem to propose a return to design-less, rapid prototyping approaches, and 'to do away' with mature methodologies such as CommonKADS. Rather, CKE proposes a change in perspective, from model-based to learning-based knowledge acquisition and knowledge system development, that can be applied to any development methodology. It further aims to direct the development of tools that can support the use of a methodology that supports CKE.

The realisation of the three principles places the development of knowledge systems under considerable stress as it requires the existence of tools and infrastructure that can support the frequent iterations, participation of the expert and recurrent deployment of the knowledge system as well as the development user-interfaces and auxiliary facilities. In this technical support is essential to make these CKE approach viable and practical. An investigation into the features that such technical support must embody is described in the remainder of this paper.

The intent of the tool support and the applications described below is to show how, using the technologies available today, the practice of knowledge engineering to can be made to embrace the dynamic and open-ended nature of knowledge and enable a learning process. This focuses initially on enabling participation of domain experts and improving the quality of the applications developed with these knowledge models.

4. A Simple Tool

The Knowledge Base Editor (KBE) is a knowledge modelling tool based on a philosophy of simplicity. It does not have an elaborated domain-model and its inference mechanism is based on simple backward chaining. The basic technology contained in the system is therefore not an elaboration in a technical sense, rather it can be perceived as a simplification of the technical possibilities that can be discerned in many other tools. This simple core is coupled with visual representation of the knowledge in the form of decision tables and extensive facilities to deploy the knowledge models. Furthermore, these deployment facilities are highly configurable and can be used to quickly and easily develop customised and dedicated user interfaces.

4.1 Key Features

The KBE has a straightforward knowledge system architecture. The knowledge is contained by a knowledge base component consisting of a domain model, a knowledge model and inference mechanism. Each of these will be described in the sections below. The difference with most other systems is that the core of the system is a component that does not contain any visualisation aspects, much like a database. The visualisation is seen as separate issue, allowing development of many different visualisations and media. This allows beyond the development of dedicated domain specific applications, including the construction of domain specific knowledge model editors. In this sense the KBE is only a default interface placed on top of this component, to allow knowledge models to be edited and maintained. The component-based approach further allows easy integration of this knowledge technology as part of a larger system. As such knowledge systems have been integrated into document information systems and GIS systems.

4.1.1 Primitive Domain Model

The domain model is formed by a set of parameters. A parameter is basically an attribute-value pair, where the value can be one of four base-types: String, Integer, Real, and Boolean. String parameters can be given a domain, which enables it to operate as an enumerated value. Boolean parameters can be configured as to what term is used to denote true and false. Each parameter further has a prompt, an explanation, a domain and a default value. The prompt and explanation are used to query the user for a value when the inference can not yield a value for the parameter. The domain can restrict the possible values that a parameter can accept or state the possible enumerated alternatives that can be chosen for the parameter.

4.1.2 Visual Knowledge Representation

	Wine type	R1	R2	R3	R4	R5
C1	Main course	red meat	poultry		fish	ELSE
C2	Turkey is served	-	[illegible]	ELSE	-	-
A1	Wine type	red	red	white	white	unknown

Figure 1 An example of a decision table

The knowledge model consists of a set of decision tables. The decision table allows visual representation of a number of related rules. The decision table has been around for some time and has been used for knowledge representation in other systems as well [16]. The left top part of the table contains the conditions, below which the actions are located. These contain references to parameters in the domain. The right side contains a tree of condition-alternatives and action-alternatives. The condition alternatives and action alternatives can contain expressions yielding a value. This can be simply a value in the domain of the parameter or a reference to another parameter. Special values include the don't-care (-), which means that the condition is not relevant and can be ignored, and the ELSE alternative, meaning the remainder of the domain.

The condition parameter is compared to the active alternatives, for example *Main course* is compare the values *red meat, poultry, fish*. The chosen value is used to determine the next set of alternatives, below the chosen alternative, as visible in the figure. If there is a don't care, the condition is skipped and the condition alternatives in the row below are chosen to be executed. When all conditions have been evaluated a column of the table's action alternatives is selected, and these actions are performed, in this case assigning a value *red* to *Wine type*. By virtue of this visual representation, it is easy to examine the contents as to their completeness and exhaustiveness. Therefore it is easy to decide whether one of the required alternatives is missing or that there are no actions associated with a particular configuration. Furthermore it is easy to add additional knowledge at any point in the decision table, as an additional condition, condition alternative, or the consequences of the table's execution can be extended with additional actions.

Some additional fineries exist but these are limited compared to this main mechanism. Backward chaining inference is easy enough to be explained to an expert and can be made more clear by showing an example consultation. The simplicity of the parameters and the visual nature of the knowledge representation allow an expert to comprehend the effect of a change to a parameter or decision table and the resulting inference well enough. In limited experiments experts have been shown to be able to enter knowledge into the system independently or with limited support from the KE.

4.2 Modelling and Application Development

Knowledge modelling can take many forms, but is mostly an incremental process that can proceed top-down or bottom-up, or using a mixed approach. The most important feature is that the minimal system that can be consulted is very small. A single decision table and a couple of parameters are sufficient for a first implementation and simple knowledge systems require no more than this. The knowledge model can be consulted at any time. This allows testing of the knowledge model to its ability to deal with a specific case. As each parameter can be used a goal, testing parts of the knowledge system, up to an individual decision table is easy. A decision table can therefore be developed incrementally as a separate functional unit. Cases can also be used to perform more general tests.

There are several ways that a knowledge model can be deployed. Standardised solutions include a stand-alone system, and a web server extension that can dish out HTML pages. Both systems operate on a series of default HTML templates, but it is possible to define a custom page for a parameter and to create elaborate pages that answer several queries at once. Dedicated systems can also be developed using knowledge-aware components that are used to create sophisticated user interfaces within standard software development environments. An example of this will be shown in the application below.

What the description shows is a simple knowledge system development environment. It is easy to understand and a knowledge model can be comprehended to its dynamic behaviour with little knowledge of information technology, i.e. usable by experts. The development of a knowledge system can be performed incrementally with ease. Furthermore it has the ability to consult the knowledge base at any moment in time, which supports verification and validation. In addition, deployment can be reasonably simple using the technologies described earlier and facilities for building dedicated systems are integral to the system's architecture. In these ways the system supports the three principles of the CKE approach.

5. Building Materials Regulation Knowledge System

The building materials regulation knowledge system (BOKS) is a knowledge based system that implements the Dutch regulations on the use of building materials[1] (BSB). BOKS is also described in more detail in another paper [17]. The BSB is a complex set of rules, which many people need to adhere to. A knowledge system was seen as an ideal way to allow people to check whether they are using building materials in conformance to the law. It is currently used by approx. 3000 people in the Netherlands, constituting a large majority of those who need to work with the BSB.

---

[1] The Dutch term is '<u>b</u>ouw<u>s</u>toffen<u>b</u>esluit', hence the abbreviation BSB. More information on the legislation can be found in [18] for those with an appreciation of the Dutch language and a fondness for legalese.

BOKS is a stand-alone computer program that contains knowledge about the BSB and is an example of a dedicated system developed using the KBE. BOKS can answer questions posed by the user, by using this knowledge and by asking questions to the user such as "what is the type of material", "where is it used" and "what are the emission values the laboratory reports". After such a consultation it gives a report that conforms to the same conclusions an expert would reach.

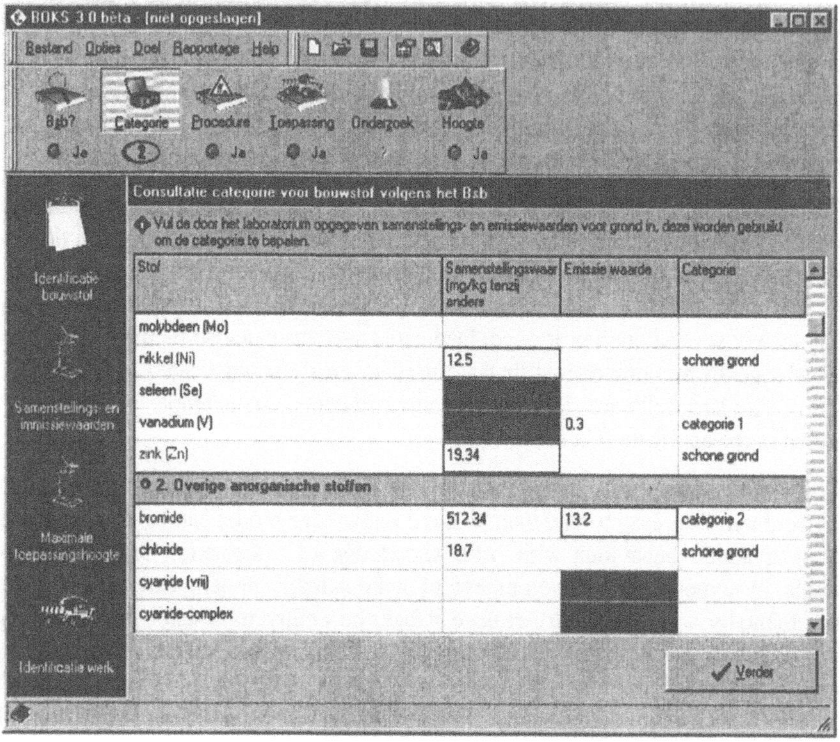

Figure 2 BOKS User Interface

The nature of all the questions asked by the system is such that a layman can answer them. The system can be consulted to answer six goal questions:

- Is the BSB applicable to the current situation,
- Which is the category the materials belong to,
- What procedures must be followed when the material is used,
- What requirements are posed on the application of the material,
- How should a sample of the material be taken and examined,
- How high is the user allowed to stack the material?

These questions offer enough information to allow a user to know they adhere to the BSB. The user is provided with an extensive report of the conclusions and reasons behind the results, and a list of the answers given to the system.

The advantages of using the system are similar to those of other knowledge systems. The benefits of BOKS compared to a human expert, make the knowledge system to be preferable in most respects, especially since there are very few human experts available. In this case the main benefit is the large number of users that would otherwise require training to acquire an intimate knowledge of the BSB, which they do not require now. It therefore negates the training of laymen in complex legislature, which they would have a hard time keeping up with, as modifications to the law are still being made. And as further proof of the effectiveness of the system, it is also being employed by the experts that have been involved in its development, because BOKS is quicker and more accurate in its findings than they themselves are.

5.1 Modelling Approach

The government publication of the BSB consists of 300 pages of text that is difficult to apply not only because of its sheer size, but also by the numerous tables, references, schemas, footnotes and exceptions. All of these have to be taken into account, and the knowledge for one specific aspect can be found spread throughout the document. Practice has shown that even experts find it difficult to use it correctly. On the other hand, once the BSB has been applied to particular situation, the results are easy to understand.

BOKS now contains about 180 decision tables and 220 parameters. During the development of the first two versions that were deployed and a third one that is currently being developed, these numbers have changed considerably (initially it contained some 120 tables), as legislation was changed or other modifications were necessary. The knowledge modelling was performed solely with the KBE. Based on the functionality of the different goals questions mentioned earlier, the system was developed by incrementally adding each of the six goals. Each goal functions as a sub-goal for other goals, therefore the development of the knowledge model in these subdivisions worked especially well.

The KE took the role of the intermediary, who would interview and discuss the global and specific parts of the BSB articles of law. Initially this was fed back to the expert as text notes. This practice was abandoned early on for an alternative propose and revise approach, where knowledge was entered by the KE in the system, and the decision tables and parameters were reviewed by the expert. This was used almost exclusively for the duration of the knowledge acquisition. This review allowed the KE to explain the content of the decision tables and perform some test cases with the system. In many cases these discussion would concern a single table, but two complete reviews of the knowledge as a whole were also performed.

The visualisation of the knowledge as decision table and the simple domain model and inference mechanism allowed the expert to not only understand it enough to validate and verify the modelled knowledge, but also to comment directly on the table, giving instructions to add an alternative or change a value in one of the cells. Sometimes this could extend to the suggestion to add a new table at a certain place. The ability to consult the system at each point in time also enabled some experimentation and the use of test cases. In other cases a single or a small set of decision tables would come under scrutiny by consulting these few tables.

Further results were that by posing an initial decision table, one could come to further clarity through iterating over the knowledge. An initial flawed formulation would enable reformulation and exploration of the actual knowledge representation that would best fit the sometimes difficult law articles. This has also led to a situation where the experts would many times faulted by their own knowledge. The consistent and complete representation of the decision table would contain a missing action alternative, indicating that a specific situation was not covered by the law. After the initial surprise the legislation would then be examined by the expert, who would try to come up with an answer.

From consultation of the system on test- and real life cases, using both release candidates and deployed versions, different errors and misconceptions surfaced. These were fed back into the model. In some cases it became obvious that these errors were caused by inconsistencies in the law itself, which have been reported to the legislative body. Furthermore, changes were made to the law in some areas. In effect the knowledge model was seen to grow considerably and the majority of the decision tables in the model have seen modification during their lifetime. The development of the system will remain to be an ongoing effort as more changes are made to the law and the system is applied to many more real world problems.

5.2 Development

The BOKS is a system that integrates different additional components. The knowledge component is only a part of the larger system; the application also contains a database of building materials, from which many of the questions otherwise posed to the user are answered, and integrates an XML-based reporting facility. The domain experts have maintained both the report templates and the content of the database.

The user interface developed for BOKS is made to be user-friendly and easy to use. Because of the simple questions it answers and the visualisation it supplies, people often suppose BOKS is no more than a friendly kind of spreadsheet. While this is the intended effect, this obscures the fact that it contains a vast array of knowledge, which it applied diligently to come to its final conclusions.

5.3 Conclusions

The application shows that the domain experts can understand the knowledge system and gain important insights into their own knowledge and, beyond this, the expert can make constructive comments directly on the knowledge represented in the system. The development of the model through small increments with the participation of the expert is perceived as critical in this respect. This gradual approach also aids the communication between the KE and the expert and makes the expert assert ownership of the model.

As to the ability for the expert involvement in the actual knowledge modelling, a final conclusion that was reached is that the expert would be able to develop parts of the knowledge model by themselves, of the granularity of a single decision table.

A problem that was discerned was the lack of structure in the domain model as a set of parameters, making it difficult to form a conception of the organisation and interconnections in the model. Locating parts of the model that either would need to be modified or that would be affected by a change in the model was made harder by this as the model grew. Some separate experiments confirm this result.

It further shows how from an initial implementation of a knowledge model, an iterative strategy can explore the knowledge, by experimentation and evolution. This improves not only the knowledge model, but also leads to significant insights for the domain expert both from the added clarity of an explicit model, revelations during modelling and from the application of the system.

6. Conclusions, Recommendations and Future Work

This paper has described an alternative approach to knowledge engineering that embraces the philosophy that knowledge systems are open-ended, dynamic artefacts that develop through a learning process in reaction to their environment. This approach is based on the debate on symbolic and situated models of human cognition and aims to make knowledge engineering a learning process by adherence to an incremental development, participation of the domain specialist and direct application of the knowledge system. This paper has presented this approach to knowledge engineering, examined a knowledge modelling tool that supports it and looked at an application of this tool.

The application and the tool make clear that the principles on which the CKE approach is based are at least attainable. The tools and techniques employed to this end are not drastically different from those employed in other knowledge system projects. This is in contrast with the RDR approach, which employs technology specifically engineered for that purpose. The application itself also shows the CKE approach is able to come to practical, industrial strength knowledge systems. Additionally, some evidence is given that the benefits that CKE proposes are also realised. The experts rather than the KE gains access to the knowledge model, and are able to improve their understanding of the domain. The technology used allows faster development of an initial system, which binds the system to its application, allowing experiences from application to feedback to the knowledge modelling. The iterative nature of the development allows gradual development of the knowledge in response to this improved understanding. This is perceived as lessening the knowledge acquisition and maintenance bottleneck.

The KBE is employed in different practical projects and further experiments with experts involved in modelling the knowledge with the tool are planned. Further development of the tool examines the role that object oriented modelling can play in creating a conceptual map of the domain, and to partition and organise the knowledge. This should alleviate most of the difficulties in locating the parts that need to be modified. It is thought that the object orientation will provide further assistance in making the models easier to build and change, through inheritance and encapsulation. This development is currently underway.

References

[1] Menzies, T.J. (1998). Towards Situated Knowledge Acquisition. In: Special Issue of the International Journal of Human Computer Studies "The Challenge of Situated Cognition for Symbolic Knowledge Based Systems", pp. 867–893.

[2] Clancey, W.J. (1999). Situated Cognition: On Human Knowledge and Computer Representations. Cambridge: Cambridge University Press.

[3] Newell, A. and Simon, H. (1976). Computer Science as Empirical Enquiry: Symbols and Search. Communications of the ACM 19(3).

[4] McDermott, J. (1989). A Critique of Pure Reason. Computational Intelligence 3, pp. 151–160.

[5] Brooks, R. (1991). Intelligence without Reason. In: Proceedings of the Twelfth International Joint Conference on Artificial Intelligence (IJCAI-91), pp. 569–595, San Francisco, CA: Morgan Kaufmann.

[6] Schreiber, G., Akkermans, H., Anjewierden, A., et al. (1999). Knowledge Engineering and Management. Cambridge: MIT Press.

[7] Bachant, J. and McDermott, J. (1984). R1 revisited: Four years in the trenches. AI Magazine 5(3), pp. 21–32.

[8] Crofts, A.E., Ciesielski, V.B., Molesworth, M., Smith, T.J., and Lee, R.Y. (1989). Bridging the Gap Between Prototype and Commercial Expert Systems – A Case Study. In: Applications of Expert Systems, J.R. Quinlan (ed.), Reading, MA: Addison-Wesley, pp. 93–105.

[9] Buchanon, B., Barstow, D., Bechtel, R. et al. (1983). Constructing an Expert System. In: Building Expert Systems, Hayes-Roth, F., Waterman, D., Lenat, D. (eds.), Reading, MA: Addison-Wesley.

[10] Compton, P., Horn, K., Quinlan, J.R., and Lazarus, L. (1989). Maintaining an Expert System. In: Applications of Expert Systems, J.R. Quinlan (ed.), Reading, MA: Addison-Wesley, pp. 366–384.

[11] Compton, P. and Jansen, R. (1990). A Philosophical Basis for Knowledge Acquisition. Knowledge Acquisition 2, pp. 241–257.

[12] Jacobson, I., Booch, G. and Rumbaugh, J. (1998). The Unified Software Development Process. Reading, MA: Addison-Wesley.

[13] Stapleton, J. (1997). DSDM: Dynamic Systems Development Method. Reading, MA: Addison-Wesley.

[14] Brooks, R. (1991). Intelligence without Representation Artificial Intelligence 47, pp. 139–159.

[15] Marques, D., Deallemagne, G., Klinker, G., McDermott, J. and Tung, D. (1992). Easy Programming: Empowering People to Build Their Own Applications. IEEE Expert June, pp. 16–29.

[16] Montelbano, M. (1973). Decision Tables. Palo Alto, CA: Science Research Associates, Inc.

[17] Spronck, P. and Schilstra, K. (2000). BOKS: A Rule-base System in Support of the Dutch Building Materials Regulations. In: Proceedings of the Sixth Pacific Rim International Conference on AI (to be published).

[18] Dutch Ministry of VROM (1998). Bouwstoffenbesluit, Parts 1, 2, and 3.

Better Knowledge Management Through Knowledge Engineering: A Case Study in Drilling Optimisation

Alun Preece, Alan Flett*, Derek Sleeman
Department of Computing Science
University of Aberdeen
Aberdeen, UK

David Curry, Nigel Meany, Phil Perry
Baker Hughes OASIS
Aberdeen, UK

Abstract: In recent years the term knowledge management has been used to describe the efforts of organisations to *capture*, *store*, and *deploy* knowledge. Most current knowledge management activities rely on database and web technology; currently, few organisations have a systematic process for capturing *knowledge*, as distinct from *data*. The paper presents a case study in which knowledge engineering practices are being used to support knowledge management by a drilling optimisation group within a large service company. The three facets of the knowledge management task are illustrated: (1) Knowledge is *captured* by a knowledge acquisition process in which a conceptual model of aspects of the company's business domain is used to guide the capture of cases. (2) Knowledge is *stored* using a knowledge representation language to codify the structured knowledge in a number of knowledge bases, which together comprise a knowledge repository. (3) Knowledge is *deployed* by running the knowledge bases within a knowledge server, accessible by on the company intranet.

1 Introduction

In recent years the term knowledge management has been used to describe the efforts of organisations to capture, store, and deploy knowledge [1, 2, 3]. Organisations are interested in acquiring knowledge from valued individuals, and analysing business activities to learn the lessons from successes and failures; such captured knowledge then needs to be made available throughout the organisation in a timely fashion. Most knowledge management activities are a combination of business processes and information technology [4]. As currently practised, knowledge management" is a broad-based activity:

- ***Document management systems*** allow workers to find existing documents relevant to the task-at-hand. Essentially these are multi-source search/information retrieval systems, tied into an organisations intranet

---

* Author's current affiliation: Interprice Technologies, Berlin, Germany.

(possibly extending to the public internet). A number of commercial products such as those of Autonomy and Verity are available.

- ***Discussion forum systems*** promote dissemination of knowledge within communities of practice. Workers subscribe to forums relevant to their interests, exchanging questions and answers, lessons learned, announcements, and industry gossip. Such systems are easily implementable with freely-available web software, in addition to commercial products.
- ***Capability management systems*** allow an organisation to "know who knows what" [5]. Essentially, they can be thought of as databases of suitably-structured CVs, and as such are implementable with off-the-shelf database software. The goal is to put people together by matching someone's need for expertise with someone else's listed skills.
- ***Lessons-learned knowledge base systems*** are designed to allow workers to tap into past experience, by storing that experience in the form of structured *cases*. These systems support sophisticated queries, typically supporting "fuzzy" retrieval of "similar" cases. While simple systems can be built using conventional database software, special-purpose case-based reasoning or knowledge-based system software is needed for full functionality.

In terms of technology, most current knowledge management activities rely on database and internet systems: if knowledge is stored explicitly at all, it is typically stored in databases either as simple tables (for example, in relational databases) or semi-structured text (for example, in Lotus Notes). There is little use of sophisticated knowledge representation systems such as Classic, Loom, or G2. Also, few organisations have a systematic process for capturing *knowledge*, as distinct from their conventional *information*-capture procedures.

This paper argues that technology and processes from the knowledge engineering field are significantly under-utilised in current knowledge management practice, despite recent efforts to promote their use [8]. Specifically, this paper focuses on:

1. use of knowledge acquisition processes to capture structured knowledge in a systematic way;
2. use of knowledge representation technology to store the knowledge, preserving important relationships that are far richer than is possible in conventional databases.

In support of this viewpoint, the paper presents a case study in which knowledge engineering practice is being used to support knowledge management by a drilling optimisation group within a large service company in the oil and gas industry. The three facets of the knowledge management task (capture, storage, and deployment) are illustrated: (1) *Knowledge capture* is performed by a systematic knowledge acquisition process in which a conceptual model of aspects of the company's business domain is used to guide the capture of cases and rules. (2) *Knowledge storage* is performed by using a knowledge representation language to codify the structured knowledge in a number of knowledge bases, which together comprise a knowledge repository. (3) *Knowledge deployment* is performed by running the knowledge bases within a knowledge server, accessible by standard web browsers on the company intranet, and capable of answering far more complex queries than is possible using conventional database systems.

The paper is organised as follows: Section 2 examines the relevant technology and processes from knowledge engineering; Section 3 introduces the application domain of the case study; Section 4 looks at the implemented system in detail; Section 5 examines the implemented system; and Section 6 concludes.

2 Applying Knowledge Engineering to Knowledge Management

In the 1990s, knowledge engineering emerged as a mature field, distinct from but closely related to software engineering [3, 6]. Chief among the distinct aspects are:

- A range of techniques for knowledge elicitation and modelling.
- A collection of formalisms for representing knowledge.
- A toolkit of mechanisms for implementing automated reasoning.

The knowledge engineering process in outline is as follows [3, 7]:

1. ***Requirements analysis***: identify the scope of the knowledge-based system, typically in terms of the competency it will be expected to have (for example, kinds of queries it will be able to answer).
2. ***Conceptual modelling***: based on the scope defined in (1), create a glossary of terminology (concepts) for the application domain and define interrelationships between the terms, and constraints on their usage. An explicit conceptual model of this kind is commonly called an *ontology*.
3. ***Knowledge base construction***: using the conceptual model/ontology from (2) as a collection of *knowledge containers* (or schemata), populate the knowledge base with *instances* of domain knowledge (often in the form of rules, facts, cases, or constraints).
4. ***Operationalisation and validation***: operationalise the knowledge base from (3) using automated reasoning mechanisms, and validate its competence against the requirements from (1). If satisfactory, release the system; otherwise, repeat (1-4) until satisfactory.
5. ***Refinement and maintenance***: after delivery, the system will continue to evolve, as knowledge changes; this will involve repeating of (1-4) throughout the life of the system.

Any knowledge management system that involves the explicit representation of knowledge is amenable to development using at least part of the above process. In fact, it can be argued that it is *always* worth applying at least part of this process when undertaking any knowledge management activity that involves the explicit representation of knowledge. For example:

- ***Document management systems***: as a minimum, apply (1) at the outset to ensure competency criteria are defined. This will ensure at least that the right tool is selected; it may reveal a need for a more structured approach.
- ***Discussion forums***: as a minimum, apply (1-2) to ensure that the scope of the system is well-understood, and that the forums are organised so as to effectively support existing (or desired) communities of practice.
- ***Capability management systems***: as above, apply (1-2) to define the metaknowledge that will serve as knowledge containers/schemata to capture workers' capabilities; then, the "CV database" is populated in step (3).

- ***Lessons-learned knowledge base systems***: these *are* knowledge-based systems, and should follow the entire 5-stage process.

It is particularly important to employ knowledge engineering techniques when an organisation seeks to employ a range of knowledge management approaches. This is becoming common in larger organisations: such organisations are already comfortable with a multiplicity of information systems, typically tied into an intranet, and see a multifaceted knowledge management system as normal. For example, such a knowledge management system may include a capability management system, discussion forums, a document management system, and several lessons-learned knowledge bases. In such cases, the key challenge becomes that of *knowledge integration*: linking together the various sources at the level of knowledge content.

In this context, the knowledge engineering process is used to define an organisational knowledge model, or *knowledge map* [9], which becomes the set of relationships that are used to bind together the multifaceted knowledge management system at the level of knowledge content. (The actual software-level bindings can be implemented by hyperlinking, remote procedure calling, or any one of a host of distributed computing techniques.) Therefore, even when an organisation embarks on its first, single-facet knowledge management project, it is very likely to be a worthwhile investment to follow steps (1-2) of the knowledge engineering process to define an initial knowledge map.

3 Case Study: Drilling Optimisation Domain

Baker Hughes OASIS is an engineering services subsidiary company of Baker Hughes Incorporated, providing drilling process expertise to a worldwide client base in the oil and gas industry. In particular, Baker Hughes OASIS specialise in *drilling performance optimisation*, which is a knowledge-rich service involving identifying, understanding, and overcoming barriers to improved drilling performance. Drilling performance optimisation engineers need a specialised set of skills, drawn from a variety of disciplines including mechanical engineering, geology, and physics. As a relatively new service, there is a limited community of skilled optimisation engineers, and those within Baker Hughes OASIS are dispersed worldwide.

For these reasons, drilling performance optimisation represents an ideal application domain for knowledge management. Having recognised this in the early 1990s, Baker Hughes OASIS have developed a multifaceted knowledge management approach, which currently includes the following systems components:

- ***Drilling Performance Guidelines***: a semi-structured document base implemented using Lotus Notes/Domino [16].
- ***OASIS University***: on-line training system for optimisation engineers, also implemented in Lotus Notes/Domino.
- ***Drill Bit Advisor***: a rule-based expert system implemented in LISP/CLOS using a custom graphical rule representation [10].

- ***Drilling Knowledge Store***: a technical lessons-learned knowledge base, described further below.

All of these components are inter-linked. For example, a conclusion (recommendation) made by the Drill Bit Advisor is commonly linked via a URL to a Drilling Performance Guideline in the Lotus Notes/Domino system.

The Drilling Knowledge Store is one of the newest components of Baker Hughes OASIS' knowledge management strategy, and has been designed as an open repository of case-based drilling knowledge. Accessed through a Lotus Domino server, its features include:

- A structured search tool that allows users to query the knowledge store for lessons learned in environments similar to a specified environment of interest.
- Forms for entry of new knowledge to promote easy entry of new cases, which are submitted to reviewers for audit and approval before being made available to other users.
- Links to the Drilling Performance Guidelines system to avoid knowledge duplication and ease updating/maintenance.

The Drilling Knowledge Store builds on a knowledge map developed using the standard knowledge engineering process described in the previous section, and incorporates a *Drilling Knowledge Repository*, which is a case-base of documented experience of optimisation engineers. The work was carried out in collaboration with the University of Aberdeen, managed as a Teaching Company Scheme. The development stages are detailed below.

3.1 Requirements Analysis

A series of initial interviews was conducted with a number of optimisation engineers to explore the scope of the Drilling Knowledge Repository. The key finding was that the system would need to be highly open: because drilling optimisation is a relatively new specialism, knowledge in the domain is evolving, so the system would have to be designed to cope with the likely kinds of change. In particular:

- any knowledge containers/schemata would have to be highly extensible (new concepts and relationships may be discovered in the future);
- instances would frequently be added (new cases will grow in proportion to the growth in the drilling optimisation business);
- instances may be reclassified, especially as outdated knowledge is "decommissioned".

3.2 Conceptual Modelling

An initial glossary of terms was drawn-up following the first round of interviewing. The transcripts of the interviews were analysed using the PC-PACK [8] knowledge acquisition software toolkit in an attempt to derive a set of concepts. However, the tool was not found to be sufficiently flexible in dealing with concepts where the "defining" words are not adjacent in a piece of text, and are interspersed with words from other concepts. PC-PACK and similar textual mark-up systems allow the user to indicate only that single words correspond to concepts, attributes, and values. In practice, it is often the case that such entities are defined by a number of words and that these are not necessarily adjacent. For example, the text "a bus system that links all the suburbs to the centre and to each other" contains the concept *comprehensive-city-bus-network*, but also contains parts of the concept *city* (*suburb* and *centre*).

In view of these limitations of the tool, a manual concept-mapping approach was used instead [11]. The conceptual modelling activity focussed on two areas:

- Defining the concepts associated with the drilling environment, including extensive definitions of geological concepts (leading to the creation of an ontology for representing the rock formations that constitute a drilling task) and concepts associated with the drilling activity (chiefly drill bits, fluids, and related apparatus).
- Defining the "knowledge management" concepts that would allow the capture of useful instances of optimisation engineers' experience (most obviously, the concept of a "case").

Relatively early on, it became desirable to formalise these concepts in order to manage them within a software environment. The Loom knowledge representation system [12] and its associated Ontosaurus browser/editor was chosen for the following reasons:

- Loom is one of the most flexible, least constraining knowledge representation systems currently available.
- Loom's operational mechanisms (chiefly the classification engine) allowed the knowledge engineering team to test the integrity of the conceptual model during its development.
- Ontosaurus provides a web front-end to Loom knowledge bases, allowing multiple users to inspect, query, and modify the knowledge base on a network, using a standard web browser.

3.3 Knowledge Base Construction

By the time a reasonably-complete conceptual model had been defined, a number of sample cases were already available, having been elicited from optimisation engineers as a natural part of exploring the scope of the domain. When formalising the conceptual model in Loom, the opportunity was taken to represent these cases using the knowledge containers therein. However, systematic acquisition of cases was carried out using a distinct approach.

Firstly, a small number of high-performing, expert optimisation engineers were identified. Then, one-on-one, intensive *knowledge acquisition campaigns* were conducted with these individuals. These campaigns were carefully designed to ensure that the experts would contribute actively and positively (mindful of lessons learned from negative experiences of knowledge acquisition during the hey-day of expert systems in the 1980s). The knowledge acquired from each campaign was formalised in Loom, but also written-up in a natural-language *knowledge book* electronic document [13] that could be checked for accuracy by the expert, and also disseminated on CD-ROM throughout the company as an easily-accessible, early result of the OASIS group's activity.

3.4 Operationalisation and Validation

The knowledge base was operationalised naturally by the choice of Loom as representation language. Validation was performed at two levels:

- Indirect validation of the represented knowledge using the knowledge books.
- Direct validation of the represented knowledge using Loom's inference mechanisms.

Further indirect validation came through the development of drill bit selection rules using some of the case-based knowledge acquired in these campaigns. These rules were then validated using existing software in the Drill Bit Advisor expert system.

The next section examines the structure of the Loom Drilling Knowledge Repository in more detail.

4 The Drilling Knowledge Repository in Loom

Philosophically, the system is considered to be a "knowledge storage and retrieval system" rather than a "knowledge-based system". The reason for this is that knowledge-based systems are strongly associated with tasks, whether those tasks are decision support, automated decision-making, training, or whatever. It is well-known that the task for which knowledge is to be used places a strong bias on the form of the knowledge [14]. In this case, it was important that the implemented system be considered task-neutral: it should serve purely as a repository for captured knowledge, without any risk of biasing the form and content of the knowledge toward a particular future usage.

Nevertheless, the system does have at its core a set of automatic deductive facilities (provided by Loom), which operate on the definitions given to the system by the knowledge modeller. The important point, however, is that these inferences operate at the conceptual model level, and not at any action level. That is, actions are left to humans, and the system does not *per se* advise on any action the user should take. So, for example, the system will recognise instances and classify them appropriately with reference to the conceptual model, but this is purely for the purpose of retrieving those instances and bringing them to the user's attention.

There are two main parts to the Loom knowledge store; a *conceptual model* part, and a *database* part. (This is analogous to a database with its schema and data parts.) The conceptual part of the knowledge base is defined using concepts. There are binary concepts (otherwise known as *roles*) and unary concepts (known simply as *concepts*). The database part is populated with *instances* of the concepts in the conceptual part.

This section gives examples of the Loom constructs to illustrate the approach taken in concrete terms. The intention here is to motivate the constructs so that a full understanding of the representation language is not necessary (if the reader is unfamiliar with Loom or similar languages, good introductions are given in [12, 15]).

4.1 Modelling Constructs for Drilling Engineers' Experience

As the knowledge store is chiefly intended to capture experiential cases from drilling engineers, the most important concept is the *case*.

```
(defconcept CASE :is-primitive
 (:and (:exactly 1 formation-sequence)
   (:all decision DECISION)
   (:all observation OBSERVATION)))
```

A case usually describes a *drill bit run* - a continuous period of drilling with a single drill bit. So, if an optimisation engineer experiences some bit run worthy of being recorded in the knowledge store, then the rock formation sequence drilled should be represented, as well as the decisions taken on how to drill that formation sequence, and any associated observations. A decision can refer to a choice of drill bit, mud (drilling fluid), flow rate, and so on. Alternatively, the case need not refer to an actual drill bit run if the person entering it simply has an experience they wish to share.

A *decision* is presupposed to have several different dimensions. These include: *issues*, *actions*, *goals*, an author, a *spin*, and *reasoning*. These dimensions are intended to provide a balance between structured knowledge and free text. The structured knowledge is to enable the formal representation and therefore support powerful searches, while the free text supports semi-structured knowledge.

```
(defconcept DECISION :is-primitive
 (:and (:exactly 1 action)
   (:at-most 10 issue)
   (:at-most 10 goal)
   (:at-most 1 authors-reasoning)
   (:at-most 1 companys-reasoning)
   (:at-most 1 author)
   (:at-most 1 spin)))
```

An *issue* is some informational context that the engineer was aware of when making the decision. The issues in the current KB reflect quite strongly the Best Practice Drilling database (held in Lotus Notes), and the link roles shown below are intended to reflect this. These can be filled with links to other media, including the Notes database itself, using URLs.

```
(defconcept ISSUE :is-primitive
 (:and KNOWLEDGE_MANAGEMENT_CONCEPT
   (:at-most 1 symptoms-and-diagnosis-link)
   (:at-most 1 description-link)
   (:at-most 1 parameters-link)
   (:at-most 1 diagnostic-information-link)
   (:at-most 1 planning-actions-link)
   (:at-most 1 operating-practices-link)
   (:at-most 1 examples-link)))
```

An *action* is the real-world consequent the engineer performed as part of the decision, and includes both structured (`categorical-outcome`) and free text (`textual-outcome`) outcomes.

```
(defconcept ACTION :is-primitive
 (:and KNOWLEDGE_MANAGEMENT_CONCEPT
   (:at-most 1 categorical-outcome)
   (:at-most 1 textual-outcome)))
```

Two kinds of *reasoning* are captured for a decision: *Author's reasoning* is a field of free-text for explanations for example of why a certain drill bit was chosen. This is to allow the possibility of incomplete, inaccurate, and even incoherent explanations for actions being stored; after all, the main reasoning or determinism for the action is the other structured information describing the circumstances in which the action was taken, such as the formation sequence. *Company's reasoning* is a field which expresses the company's commonly agreed upon beliefs for the decision in question.

4.2 Modelling Constructs for the Drilling Environment

The drilling environment is described chiefly in terms of conceptual rock sequences. Representing these was achieved by defining an ontology of geological concepts, including constraints. For instance, if the user wishes to specify the depth and/or length of a particular section of *lithology* (a basic rock type, for example, sand, shale, etc) then that section has to be represented as a *formation.* The super-structure larger than that is the *formation sequence*, which can have one or more *formations.* Each formation can have one or more *lithologies.* A formation is the conceptual modelling granularity at which the users should be representing any part of the wells they feel should have represented interval lengths and depths.

```
(defconcept FORMATION_SEQUENCE :is-primitive
 (:and ROCK_CONCEPT
    (:at-least 1 formation)))

(defconcept FORMATION :is-primitive
 (:and ROCK_CONCEPT
    (:at-least 1 lithology)))
```

To allow users to represent and query formation sequences in a flexible way, a number of relations are defined in the ontology. For example, the relation `comes-in-somewhere-after` relates two formations, the first of which comes in somewhere after the other.

```
(defrelation comes-in-somewhere-after
 :domain FORMATION_SEQUENCE
 :range FORMATION
 :characteristics (:multiple-valued :closed-world)
 :is (:satisfies (?formation-x ?formation-y)
          (:and
           (FORMATION ?formation-x)
           (FORMATION ?formation-y)
           (:or (comes-in-immediately-after
                  ?formation-x ?formation-y)
             (:exists (?formation-z)
                   (:and
                    (FORMATION ?formation-z)
                    (comes-in-somewhere-after
                       ?formation-x ?formation-z)
                    (comes-in-somewhere-after ?formation-z
                           ?formation-y)))))))
```

One important feature of *lithologies* in their *hardness.* While a lithology has, by definition, one rock type (for example, shale), it can have more than one hardness (for example, that shale may have 100m of *very soft rock* with 300m *soft rock*).

```
(defrelation hardness
 :domain LITHOLOGY
 :range HARDNESS
 :characteristics (:closed-world :multiple-valued))
```

The ontology has a collection of functions that relate formation sequences, constituent lithologies, and accumulated hardness, to support the modelling of drill bit runs.

In addition to the generic geological concepts, the knowledge store has representations of the concepts involved in *drilling*. For example, *drill bit*.

```
(defconcept DRILL_BIT :is-primitive
 (:and DOWN-HOLE_EQUIPMENT_CONCEPT
   (:exactly 1 bit-gauge)))
```

4.3 Querying the Knowledge Store

The function `retrieve` provides an interface to Loom's deductive query facility, used for retrieving instances from the knowledge base. *Formation sequence queries* are among the most sophisticated forms of query that can be issued to the knowledge store. The two concepts likely to be of interest are individual formations, and formation sequence. Two common forms of query here are:

- Queries on an overall cumulative amount of a certain hardness of a certain lithology over a formation sequence.
- Queries for formations that have amounts of certain lithologies of certain hardness.

The example query below looks for cases that have a formation sequence which has as its constituents of its formation(s) greater than or equal to 1900 feet of very soft to soft (including all subtypes of soft and very soft) shale (including all sub-types of shale).

```
(retrieve ?case
  (:and
   (CASE ?case)
   (>= (sum (:collect ?lithology-amount-ft
              (:and
               (:exists (?formation-sequence ?formation
                     ?lithology ?hardness)
                 (:and
                  (formation-sequence ?case ?formation-sequence)
                  (formation ?formation-sequence ?formation)
                  (lithology ?formation ?lithology)
                  (lithology-hardness-amount-ft ?lithology
                     ?hardness ?lithology-amount-ft)
                  (:or
                   (VERY_SOFT ?hardness)
                   (SOFT ?hardness))
                  (SHALE ?lithology)
                  ))))) 1900)))
```

Users are also likely to want to look for cases where specific goals (outcomes) were achieved. The following example query retrieves cases that have a drill bit decision in which one of its goals was *good ROP* (rate of penetration) with *good bit cleaning*.

```
(retrieve ?case
    (:and
     (CASE ?case)
     (:exists (?decision)
          (decision ?case ?decision)
          (DRILL_BIT_PLANNING_DECISION ?decision)
          (goal ?decision GOOD_ROP_WITH_GOOD_BIT_CLEANING))))
```

4.4 Adding to the Knowledge Store

Recall that the knowledge store is comprised of a conceptual part and and a database part. It is expected that the conceptual part is now stable, and it is rare that knowledge will need to be added or modified. However, it is expected that additions to the database part will be regular. The Loom operations used to update the database part of the knowledge base are `tell` and `about`: `tell` is used to assert propositions and facts about the world or domain; `about` references the instance to which those propositions refer. The following example shows how a case instance may be entered. This example case has one formation sequence name and zero or more decisions/observations.

```
(tell (:about Case-Name
        CASE
        (formation-sequence Formation-Sequence-Name)
        (decision Decision-Name)
        (observation Observation-Name)))
```

5 Current Status and Future Plans

The Loom Drilling Knowledge Repository currently contains 1200 concepts and 240 relations, with further expansion planned. The knowledge store is accessible on the company's intranet using a standard web browser via the Ontosaurus system - see Figure 1. While it is relatively straightforward to browse the case base and ontology using Ontosaurus, there are a number of significant problems with the current system:

- it is hard to run complex queries on the Loom knowledge store because Ontosaurus supports only simple queries (retrieve cases with matching simple role values);
- it is hard to add new cases because these require the user to have good knowledge of Loom syntax, which is an unrealistic expectation for optimisation engineers;
- limited capability for multi-user access (crude database locking and limited concurrent access).

In addition to the above issues, it was desirable that the Drilling Knowledge Repository have a familiar interface, preferably that of the existing systems implemented using Lotus Notes/Domino. At the same time it was thought valuable to link the knowledge represented in the Loom repository with information and knowledge relating to the performance optimisation projects during the course of which the stored knowledge was created. For these reasons, it was decided as an interim solution to incorporate (partially) the Loom knowledge map and cases into a Lotus Notes/Domino database of project-related knowledge, to provide structure for technical lessons learnt on each project. A screenshot of the ported system appears in Figure 2. The immediate benefits of this were:

- easy access to the most valuable knowledge in the knowledge store to all optimisation engineers;
- linkage between optimisation-related knowledge and other knowledge from the projects in which it was created;

- familiar interface, seamlessly integrated with other Notes/Domino knowledge sources;
- Domino's scalable architecture, with built-in support for concurrent access.

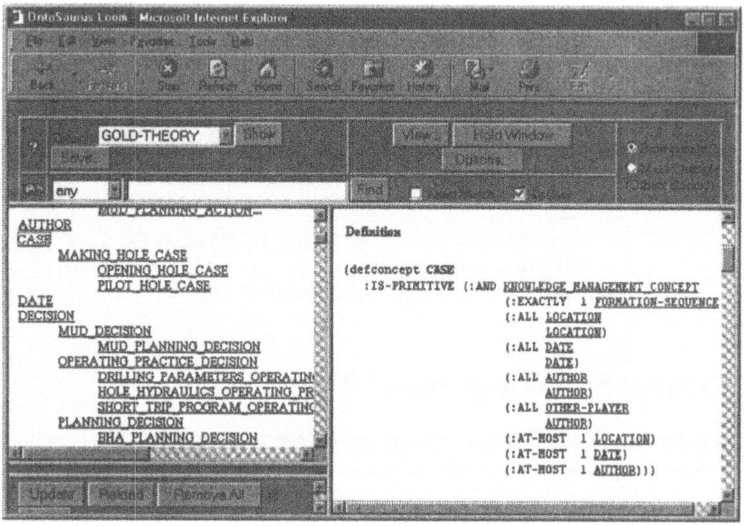

Figure 1: Loom Drilling Knowledge Repository Screenshot

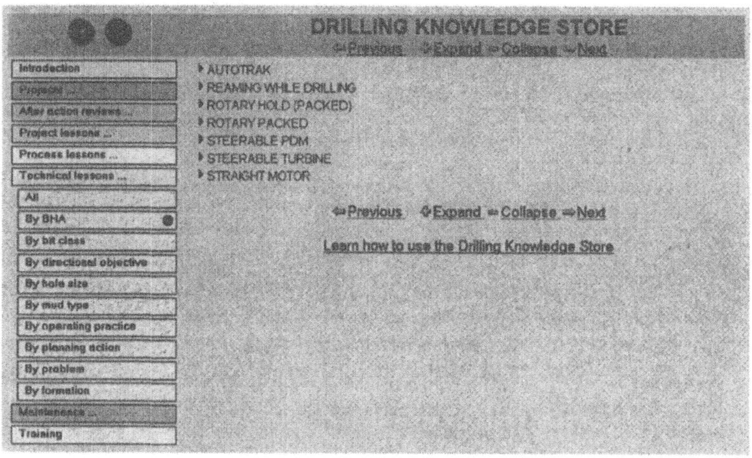

Figure 2: Lotus Notes/Domino Drilling Knowledge Store Screenshot

The port of the Loom knowledge map and cases to Lotus Notes/Domino was carried out manually. Moreover, the Notes knowledge schemata are nowhere near as rich as those in the Loom system. As this is obviously not ideal, future work will focus on providing an automatic conduit for knowledge exchange between the Notes/Domino and Loom systems, allowing knowledge management engineers to maintain the knowledge map through the Loom system, and optimisation engineers to retrieve and enter cases via the Notes/Domino system.

6 Conclusions

This paper has argued that knowledge engineering techniques bring significant benefits to knowledge management projects. The case study in drilling optimisation reveals three major areas of specific benefit to the knowledge management group at Baker Hughes OASIS:

- The *use of a principled methodology for knowledge acquisition* lead to effective capture of valuable knowledge, and instilled in all staff concerned a good degree of confidence in the process.
- The *development of a common ontology* as a result of the knowledge modelling process had benefit not only for the specific task of acquiring experiential cases from company experts, but also provided a knowledge map with application to multiple company knowledge sources.[1]
- While semi-structured knowledge books were used to provide feedback to participating personnel, the *formal representation of acquired knowledge in an operationalisable form* was very valuable for knowledge verification and integrity checking.

In doing this work, we have detected two weaknesses in current knowledge engineering techniques and technology. Firstly, as noted in Section 3.2, PC-PACK and other textual mark-up systems do not cope adequately with concepts that are defined by a number of non-adjacent words. Thus, we have identified the need for a more flexible tool.

Secondly, it is very difficult to integrate expressive reasoning tools such as Loom with intranet knowledge management environments like Lotus Notes/Domino. It seems reasonable to conclude, therefore, that while knowledge engineering processes are ready to bring significant benefits to knowledge management projects, the knowledge engineering toolbox needs some improvement.

Acknowledgements

The case study described in this paper was supported by Teaching Company Scheme funding from the DTI and EPSRC. The Lotus Notes version of the Drilling Knowledge Store was constructed largely by John Lawton and David Bowden of Transition Associates, who also proposed and developed the OASIS University. The authors would like to thank Tom Russ of ISI/University Southern California, for advice and technical assistance regarding Loom and Ontosaurus.

This paper is published with the kind consent of the Hughes Christensen Company.

[1] We have subsequently developed a knowledge acquisition tool that is driven by the nature of the task and the ontology. We believe that this tool, COCKATOO [17] would be effective in acquiring cases directly from domain experts.

References

1. *Harvard Business Review on Knowledge Management*, Harvard Business School Press, 1998.
2. Liebowitz, J. and Wilcox, L. (1997) *Knowledge Management and Its Integrative Elements*, CRC Press.
3. Schreiber, G., de Hoog, R., Akkermans, H., Anjewierden, A., Shadbolt, N. and van de Velde, W. (2000) *Knowledge Engineering and Management*, MIT Press.
4. Bukowitz, W. and Williams, R. (1999) *Knowledge Management Fieldbook*, Prentice-Hall.
5. Stader, J. and Macintosh, A. (1999) Capability modelling and knowledge management. In *Applications and Innovations in Intelligent Systems VII*, Springer-Verlag, pp 33-50.
6. Motta, E. (1999) *Reusable Components for Knowledge Modelling*, IOS Press.
7. Russell, S. and Norvig, P. (1995) *Artificial Intelligence*, Prentice-Hall.
8. Milton, N., Shadbolt, N., Cottam, H. and Hammersley, M. (1999) Towards a knowledge technology for knowledge management. *International Journal of Human-Computer Studies*, 51:3 615-641.
9. Domingue, J. and Motta, E. (2000) Planet-Onto: from news publishing to integrated knowledge management support. *IEEE Intelligent Systems*, May/June, pp. 26-32.
10. Evans, J. M., Fear, M. J., and Meany, N. C., (1995) A new graphical representation for rule definition and explanation in an expert system. In *Applications and Innovations in Intelligent Systems III*, Springer-Verlag.
11. Kremer, R. (1997) Concept mapping tool to handle multiple formalisms, In *AAAI Spring Symposium on Artificial Intelligence in Knowledge Management*, AAAI Press.
12. MacGregor, R. (1991) The evolving technology of classification-based knowledge representation systems. In *Principles of Semantic Networks: Explorations in the Representation of Knowledge*, Morgan Kaufmann, pp 385-400.
13. Ermine, J-L. (1998) Knowledge management in the Commissariat a l'Energie Atomique. In *PAKeM98 - Practical Application of Knowledge Management*, London.
14. Clancey, W. J. (1992) Model construction operators. *Artificial Intelligence*, 53:1-115.
15. Brachman, R. J., McGuinness, D. L. Patel-Schneider, P. F., Resnick, L. and Borgida, A. (1991) Living with Classic: when and how to use a KL-ONE-like language. In *Principles of Semantic Networks: Explorations in the Representation of Knowledge*, Morgan Kaufmann, pp 401-456.
16. Curry, D. A., Singelstad, A. V., and Bowden, D. (1999) Drilling Performance Guidelines - A tool for sharing drilling-related knowledge and experience", SPE/IADC paper no 52804, *SPE/IADC Drilling Conference*, Amsterdam.
17. S White (2000) *Enhancing Knowledge Acquisition with Constraint Technology*, PhD Thesis, The Department of Computing Science, University of Aberdeen.

An Integrated AI Package for Modelling Applications

W. Dixon, Q. Mehdi, N. Gough and J. Pitchford
School of Computing and Information Technology
University of Wolverhampton, Wolverhampton, England
WV1 1SB
E-Mail: CM1815@wlv.ac.uk

Abstract

This paper details the production of a software package that provides an interface to Artificial Intelligence (AI) software in order to afford modelling of systems using Adaptive Neuro-Fuzzy Inference System (ANFIS) technique, multiple linear regression (MLR), and generalised regression neural networks (GRNN). The interface allows the rapid production of models without the need for knowledge of the function of the AI software or of the techniques being used. Information on the AI techniques is provided through the package but modelling and comparison of models may be performed without accessing the information.

The modelling application is that of thermally induced errors in Computer Numeric Controlled (CNC) machines, which cause inaccuracy in the components produced. Absolute measurement of positional errors is difficult during machining but it has been found possible to derive positional error from temperature measurements. The package produces models of thermally induced errors from temperature measurements that can be used for error compensation. The package allows models to compensate for these errors to be produced and evaluated by users who have little or no knowledge of the techniques used.

Keywords: Artificial Intelligence, Neuro-Fuzzy, neural networks multimedia, CNC machines

1. Introduction

Systems may be modelled using a variety of techniques of which ANFIS, GRNNs and MLR are examples. The Multimedia & Intelligent Systems Research Group (MIST) is investigating the development and integration of multimedia and Artificial Intelligence (AI) tools to facilitate system modelling. The resulting package is referred to as Multimedia Intelligent Machine Tool Assistant (MIMTA) which has been directed towards the compensation of thermally induced errors in

CNC machine tools. MIMTA could be used to produce a model of any system and is not restricted to the thermal error application. This paper introduces a multimedia package that affords system-modelling facilities together with information on the modelling techniques used.
The modelling process can be carried out using the AI software, in this case Matlab, directly. This requires knowledge of the AI techniques and also of the Matlab package. The MIMTA package effectively hides these technicalities from the user.
The paper is organised as follows; section two introduces the modelling techniques available and typical applications; the problem addressed here, that of thermally induced errors in CNC machine tools, is discussed in section three; the software package is presented in section four; conclusions are in section five.

2. Modelling techniques

A brief overview of the techniques that have been used follows. The techniques are used to model the dependant variable (in this case the thermally induced error) from the independent variables (the temperature measurements on the CNC machine).

2.1 Multiple Linear Regression

MLR is a technique for deriving dependant variables from independent variables, producing constants which when multiplied by the independent variables approximate the dependant variable. Equation 1 shows the empirical formula used:

$$\text{Dependant variable} = b_0+b_1t_1+b_2t_2..+b_nt_n \qquad (1)$$

where $(b_{0,\dots,}b_n)$ are constants and $(t_{0,..}t_n)$ are the independent variables. The constants may be found by the use of multiple linear regression using the equation shown as equation 2 (for the case of two independent variables)[1].

$$\underbrace{\begin{bmatrix} N & \sum_{i=1}^{N} x1i & \sum_{i=1}^{N} x2i \\ \sum_{i=1}^{N} x1i & \sum_{i=1}^{N} x1^2i & \sum_{i=1}^{N} x1ix2i \\ \sum_{i=1}^{N} x2i & \sum_{i=1}^{N} x1ix2i & \sum_{i=1}^{N} x2^2i \end{bmatrix}}_{X} \underbrace{\begin{bmatrix} b0 \\ b1 \\ b2 \end{bmatrix}}_{B} = \underbrace{\begin{bmatrix} \sum_{i=1}^{N} yi \\ \sum_{i=1}^{N} x1iyi \\ \sum_{i=1}^{N} x2iyi \end{bmatrix}}_{Y} \qquad (2)$$

Where N = number of samples, y= dependant variable, x1 and x2 are independent variables and b0, b1, b2 are the constants generated.
The data for the matrices X and Y are derived from the experimental data according to the formulae indicated in the matrices. The matrix B containing the constants is generated from the data by:

$$B = X^{-1}Y$$

The constants in B may then be used to generate output data from a new set of input data; the validity of the results may be checked by comparing them with the axis deviation data.
MLR has been used to model thermal errors from temperature measurements for compensation purposes [2,3,4]. The method is effective, but it is necessary to identify the most useful points on the CNC machine to place the temperature sensors [5,6]. In order to provide exact compensation the temperature distribution over the whole machine should be known [7]; this will result in increased cost. The ideal is to find the best model produced with the smallest number of sensors, the CNC controllers in use in the current work have 2 channels of analogue to digital conversion (A/D) which implies that the best model possible from only 2 temperature measurements is the goal.

2.2 ANFIS

Adaptive Neuro-Fuzzy Inference System (ANFIS) is a method of tuning a fuzzy inference system (FIS) using a backpropagation algorithm based on input and output data [8,9].
FIS rules and membership functions may be created by an expert with knowledge of the behaviour of the system, but it is possible to generate the FIS automatically from the known input and output data together with the specification of the number and types of membership functions. The number and type of membership functions may be found by trial and error. The MATLAB command genfis1 is used to generate a FIS from the observed data [9]. The command takes as inputs the training data set, a matrix composed of the input and output data It is necessary to specify the number and type of membership functions (MF) to be used in the FIS. The MATLAB command ANFIS is then used to tune the FIS [9,10]. It is necessary to determine the number and type of MFs that give the best model by trial and error.
In order to assess the efficiency of the method the output data from the training data may be plotted together with the output produced from the trained FIS when the training data input set is used. The degree of difference between the known and predicted outputs may then be assessed and this may be used to assess the usefulness of the ANFIS approach for modelling. The production and training of a FIS using the ANFIS method is covered in [9], pp 2-63 to 2-71.

2.3 Generalised Regression Networks

Neural networks (NNWs) are based upon the neurones which constitute the human brain, each unit performing a simple operation, but a large number of interconnected neurones being capable of very complex behaviour. The processing ability is stored in the weights of the interconnections between neurones, which are set by learning from training data [11]. NNWs have been used for thermal error compensation [12-16] and for control of cutting force during machining [17].
Generalised regression networks (GRNNs) are often used for function approximation and consist of a radial basis layer and a linear layer [18]. A radial

basis function (RBF) has an output, which falls off in a circularly symmetric fashion as the input differs from the template defined from the weights. The output from the node is therefore a function of the degree to which an input matches the inputs used to create the node. The number of nodes in the first layer is equal to the number of input/target vectors in the training data used to produce the network, a node being produced for each set of data in the training set. The first layer of the radial basis network consists of a series of RBFs, the second (output) layer is made up of linear units that compute the weighted sum of the RBF units. The output of the radial basis network will depend upon the degree of match between the input data and the training data.

In GRNNs the second layer is slightly different. It also has as many nodes as input/target vectors. The output of one of the RBF functions will be close to one if the input data is a close match to the training data that produced that node. This in turn leads to a layer 2 output close to the target, which produced the node.

The Matlab Neural Network Toolbox was used for the production of GRNNs to model thermally induced errors. The function *newgrnn* is used to produce the network, separate networks being produced for each axis.

```
Net = newgrnn(P,T)
```

Where P contains the input vectors and T contains the target vectors. This Matlab function then generates the GRNN that simulates the training data.

The network may then be evaluated using

```
Out = sim(net,P)
```

where P contains the checking data input vectors. The vector *Out* will then contain the modelled output data, which can be compared with the observed data to allow evaluation of the model.

GRNNs have been used to model thermal error in CNC machines [19]. They have been found to produce slightly inferior results to those of ANFIS and MLR but the technique is included, as the effectiveness of GRNN may be better for modelling of other systems.

3. Thermally Induced Errors in CNC Machines

The architecture and operation of CNC machines has been discussed previously [10]. During CNC machining heat is which results in distortion of the CNC machine. Thermal distortion of CNC machines results in loss of accuracy and may account for 40-70% of the errors in components produced [20]. As the machine is computer controlled accuracy may be maintained by addition of offsets to the co-ordinates used by the program. These offsets may be found by direct measurement of the errors (which is expensive to perform) or by using a model to predict the errors from easily measured machine parameters, in this case the temperature at various points on the machine.

Figure 1 shows the deviation of the axes of a vertical machining centre under test. The test procedure and experimental data have been described previously [21]. The test is not wholly representative of cutting conditions as the machine axes are not moved, but the heat produced at the spindle bearings is believed to be the major source of heat input to the machine tool structure [22].

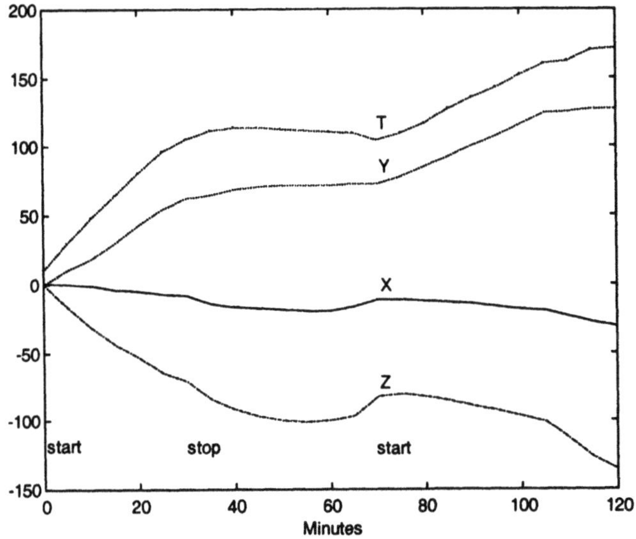

Figure 1 Axis Deviation during test

The temperature (T) during the test is also shown in Figure 1, which is (temperature-20) * 10 in order for the correlation between temperature and axis deviation to be seen. The axis deviations are shown in microns. The stated aim of CNC manufacturers is to reduce the thermally induced error below 10 microns; the uncompensated machine shows errors greatly in excess of this.

4. Software Package

The software package was developed to integrate multimedia information on AI techniques, multimedia and CNC machines together with an interface to allow the production of system models. The package is divided into two main components, the information component and the modelling component.

The information component provides access to recent literature and hyperlinks to the topics covered. A search facility is also provided for more up to date information. The modelling component affords access to previously evaluated techniques for the modelling of thermal errors, enabling the rapid production and evaluation of models for use in thermal compensation. A schematic of the package is shown in Figure 2.

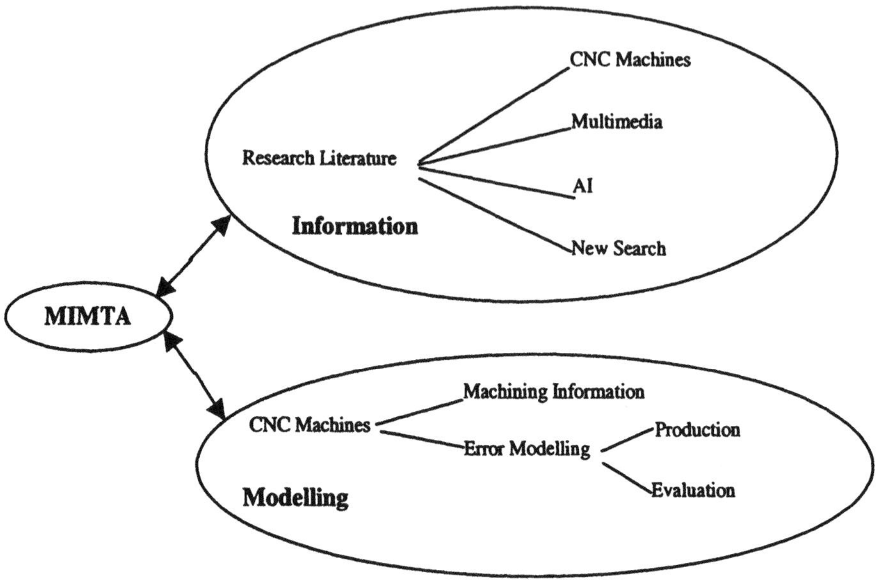

Figure 2 Structure of MIMTA

4.1 Software Development

The information and multimedia facilities and GUI component of the package was produced using Microsoft Visual Basic (VB). The AI and maths engine used was Matlab.

Communication between the interface and Matlab was achieved by two methods:

I. The Matlab engine may be started from the VB program by using the VB *shell* command, which is used to start another process from VB. The command

```
Returnvalue = Shell ("c:\matlab\bin\matlab",vbHide)
```

may be used to start the Matlab engine, the window will be active but hidden. Commands and data may be sent to Matlab as strings, using the *SendKeys* command [23]. Commands may then be built within the VB program from input supplied by the user and sent to Matlab to be executed. For example, assuming that TrainDataFiles is a (user input) variable which contains the names of the data files to be used in production of a model of thermal error

```
SendKeys "fid = fopen{(}'" & TrainDataFiles(1) & "','r'{)};~",True
SendKeys "Input{(}:,1{)} = fscanf{(}fid{)},{%}f\n'{)};~",True
```

SendKeys "fclose{(}fid{)};~",True

will result in the Matlab variable Input(:,1) holding the data from the file TrainDataFiles(1) from VB. Data may be loaded in this fashion into Matlab variables following which a predefined Matlab script file may be executed to produce a thermal error model and display the results. Predefined Matlab scripts are written to produce models using parameters which have been found to be the most suitable during previous investigation. Execution of a Matlab script (or any Matlab command) may be invoked from Visual Basic by using SendKeys as before.

SendKeys "ScriptName~"

The format of the SendKeys command is quite complex, some chacters have special meaning [23]. The variable set true using

SendKeys command,True

is the Wait mode, if set true control is not returned to VB until the Matlab command has been executed, an important consideration if commands are being executed sequentially.

II. The Matlab engine may be started from VB by using the *CreateObject* command [23], which is the method of starting any application which supports automation. Matlab may be started using

Set matlab = CreateObject("matlab.application")

Matlab commands, whether predefined or constructed from user input to VB may then be executed using

Res = matlab.execute(command)

The commands may be built from string and numeric variables. For example, assuming the variable x has already been input to Matlab, and z is a VB variable.

Res = matlab.execute("y=x*" & z)

Dynamic Data Exchange [24] may be used to exchange variables between VB and Matlab. The commands *putfullmatrix* and *getfullmatrix* will send variables to and retrieve variables from Matlab. The only complication is that all numbers are viewed as complex matrices. The following code was used to return a single value from a Matlab application to VB.

Dim RealError(0, 0) As Double *real part of error*
Dim ImaginaryError(0, 0) As Double *imaginary part of error*

Set matlab = CreateObject("matlab.application") *start matlab*

MaxError = matlab.execute("findmaxerror") *gives MaxError as a string*

Call matlab.getfullmatrix("maxerror", "base", RealError, ImaginaryError) *MaxError as a matrix* (maxerror is the Matlab variable, base is the section of Matlab memory holding it)

MaximumError = (RealError(0, 0)) *The real part of maxerror (there is no imaginary part)*

Dimensioning the real and imaginary parts of the error as doubles rather than arrays of doubles results in a run time error, as matrices are always expected.
The use of DDE for exchanging data to and from Matlab allows the presentation of results by the more friendly VB interface rather than as a line of text in a Matlab screen.

4.2 Information Component

The opening screen of the MIMTA package is shown in Figure 3.

Figure 3 MIMTA Opening Screen

Three menus are available on the screen:

i. The help menu gives access to basic information on the use of the package.

ii. The options menu has the choice to switch the Microsoft Agent, which is the parrot perched on the CNC controller, on or off. The agent can perform various animations (provided by Microsoft and invoked by the programmer) and can also speak text. The animations can catch the attention of the user, which can be useful to ensure that some vital piece of information is noticed (for example if during modelling some previously determined tolerance is exceeded).

iii. The sound menu controls what if any sounds are produced by the program. These fall into two categories, noises which are produced on for example clicking on an option, and speech which can exist on different levels of detail. If the Microsoft Agent is on speech will be performed through it; otherwise speech may be performed through a speech engine. The noises may be switched on or off, speech may be fully on or off or restricted by category (for example warnings only).

Three options are available on the opening screen:

- The CNC Machines option opens the CNC (Figure 4) screen which contains links to CNC sites of interest, which provide both background information on machining and machines as well as recent research information. The modelling interface is also accessed from this option.

- The recent literature option provides access to abstracts of recent papers, and the facility to search for recent papers, in the fields of AI, CNC and Multimedia. The abstracts of recent papers have been obtained by previous literature searches and give an up to date view of research being performed in the areas covered. Further links to internet sites are provided on these pages

- The exit option closes the program.

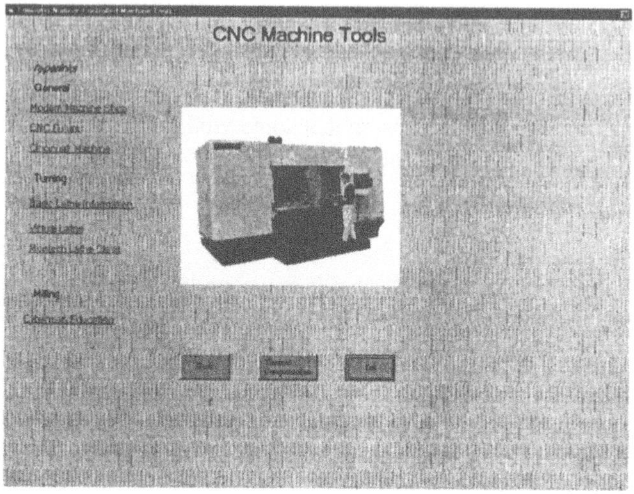

Figure 4 CNC Screen

4.2 Modelling Component

The thermal compensation option may be chosen from the CNC screen to perform modelling, this opens the Thermal Compensation screen. The menus on the thermal compensation screen afford the following options:

i. Load, this option allows the loading of data to produce or evaluate models
ii. Run. This has options to run MLR, ANFIS or GRNN modelling on the loaded data set, or to run and compare ANFIS and MLR modelling.
iii. View, allowing the test data to be viewed graphically.

The first operation is to load the data from which the model is to be produced as shown in Figure 5. The input (independent) variables are requested first, then the dependent variable.

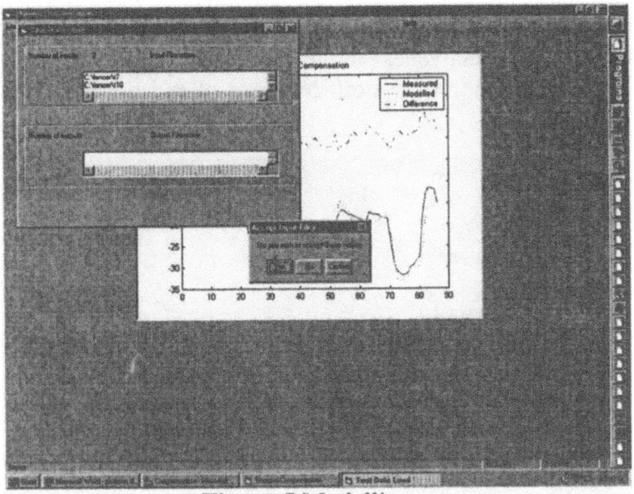

Figure 5 Modelling

Once the data has been loaded modelling may be performed. The Run menu contains the various modelling techniques. ANFIS, MLR or GRNN models may be chosen, following which the model will be produced and the results of using the model to reproduce the data used in its production are shown graphically, as shown in Figure 6 for a GRNN model. The models are produced by sending the names of the data files to the Matlab engine and then executing predefined Matlab scripts to produce the specified model using those parameters which were previously found to be most successful.

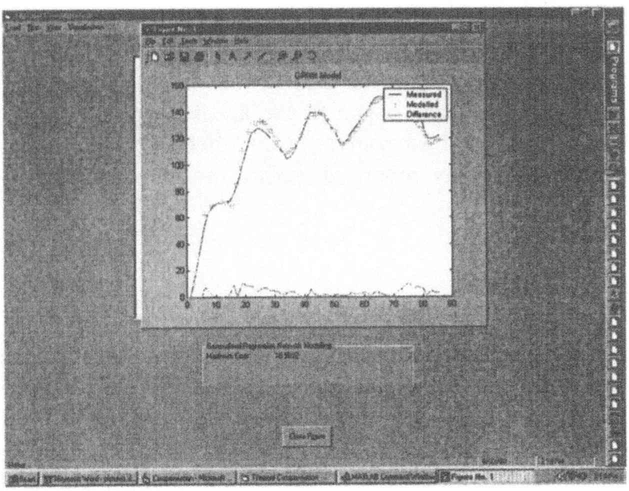

Figure 6 Model Evaluation

The user is given the opportunity to save any model produced. The compare option in the run menu will produce both MLR and ANFIS models of the system and give a comparison of their effectiveness in the form of a graph and textual information. An animated demonstration of the effectiveness of the model is also available as shown in Figure 7.

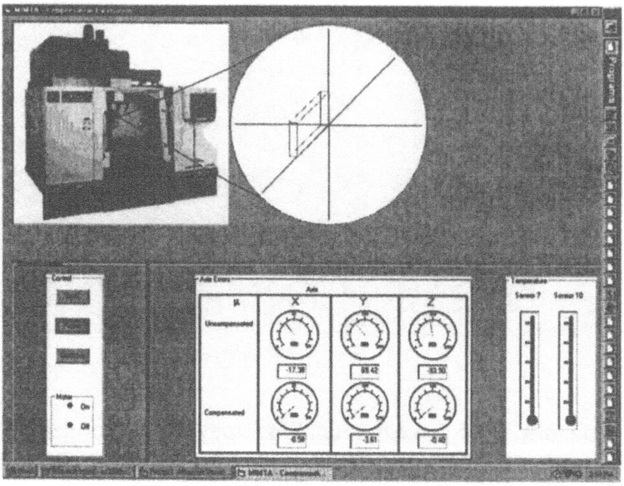

Figure 7 Animation of Model Performance

Here the temperature data is input to the model sequentially and is displayed on the temperature display as a thermometer. The observed axis deviation for each axis is displayed on dials, the compensated deviation (i.e. that resulting from subtracting the predicted deviation from the observed) being displayed below. The circular window gives a representation of the axis error display as a volume (the axis error is referred to as *volumetric error*). All display components are produced using ActiveX methods. In the event of the desired maximum (10 μ) being exceeded a warning will be displayed and spoken (if the agent or speech engine are active). The animation gives a quick assessment of the effectiveness of the compensation model; the graphical display produced during model production gives a more detailed view.

5. Conclusions

The production of an integrated package combining multimedia presentation of information together with AI modelling techniques has been demonstrated. The package simplifies the use of AI software by inexperienced users to model systems, making it possible for non-experts to utilise AI techniques. The interface removes the need to understand the techniques used from the production of thermal error models from test data for compensation purposes. Visual and sound aids are used to simplify the production of models, knowledge of the location of the input and output data files is all that is required.

The information component of the package is of necessity linked to the problem under consideration. The modelling component could be applied to any system, with changes only in the presentation of information. The package combines the latest information on the topics with the facility to perform further searches and integrates the topics to provide a multimedia front end for system modelling.
The package has given models of thermal error on CNC machines, which have proved valuable in error compensation [10,19,21], models have been produced which provide a significant improvement over regression. Models can be produced using regression, ANFIS and GRNN and compared automatically to aid the selection of the best model. MIMTA can be used to model other systems.
Further work will include investigation of new techniques for modelling thermal errors. If other techniques are found to be useful for production of thermal error models they will be incorporated into MIMTA. User-friendly interfaces for the investigation of other problem areas may be investigated, the areas of lathe control [17] and image analysis are under consideration.

Acknowledgement

The authors wish to thank Cincinnati Machine U.K. Ltd. for their participation and support during the course of this work.

References

1 Walpole RE, Myers RH. Probability and Statistics for Engineers and Scientists. Prentice-Hall, 1985

2 Hardwick BR. 1992, Improving the Accuracy of CNC machine Tools using Software Compensation for Thermally Induced Errors. In: Proc 29th Int MATADOR Conference.1992, pp269-272

3 Kurtoglu, A., 1990, "The Accuracy Improvement of Machine Tools," *Annals of the CIRP*, Vol. 39, No. 1, pp. 417–419

4 Hardwick B.R., 1993 "Further development of techniques for software compensation of thermally induced errors on CNC machine tools", *Proc Int Conference on Laser Metrology and Machine Performance (LAMDAMAP 93)* 47-63 ISBN 1-85312-241-6 Computational Mechanics Publications Southampton.

5 Spur G, Hoffmann E, Paluncic Z, Benzinger K, Nymoen H. Thermal Behaviour Optimisation of Machine Tools. Annals of the CIRP 1988; 37: part 1 401-405

6 Weck, M., Schulze, 0., Michels, F., Bonse, R.:1994: Optimization of Machine Tool Performance and Accuracy, ASME Annual Winter Meeting, October 1994

7 Donmez MA., Blomquist DS, Hocken RJ, Liu CR, Barash MM. A General Methodology for Machine Tool Accuracy Enhancement by Error Compensation. Precision Engineering; 1986a

8 Jang Roger J-S. ANFIS: Adaptive-Network-Based Fuzzy Inference System. Transactions on Systems, Man, and Cybernetics 1993; 23: No 3 665-685

9 Jang Roger J-S, Gulley N. Fuzzy Logic Toolbox for Use with MATLAB. The MATHWORKS Inc., 1995

10 Dixon, W., Mehdi, Q.H., Gough, N.E. and Pitchford, J., 1999b, "Neuro-Fuzzy and Regression Techniques for CNC Thermal Error Compensation", *Proc. 19th SGES*

Int. Conf. on K.B.S. and Appl. A.I.: ES99, December 1999, Cambridge University, UK., 304-322

11 Gurney, K, 1997, "An Introduction to Neural Networks", UCL Press Limited, London

12 Hatamura, Y., Nagao, T., Mitsuishi, M., Kato, K., Taguchi, S., Okumura J., Nakagawa; G., Sugihita H.: 1993, Development of an Intelligent Machining Centre Incorporating Active Compensation for Thermal Distortion, *Annals of the CIRP* Vol. 42/1/1993, pp. 549-552

13 Yang, M., and Lee, J., 1998, "Measurement and prediction of thermal errors of a CNC machining Centre using two spherical balls", *Journal of Materials Processing Technology,* Vol 75 190-189

14 Moriwaki, T. and Zhao, C., 1992, "Neural Network Approach to Identify Thermal Deformation of a Machining Center", *Proc. of 8th Int. IFIP TC5-WG5.3 Conf.(PROLAMAT)*, 685-697.

15 Dehaes, J., Kleiwas, P., Nuttin, M., Vanherck, P.:1995, Compensation of Thermal Deformations in Machine Tools with Neural Nets, Proc Second International Workshop on Learning in Intelligent, Manufacturing Systems, Budapest, 1995, 746-756

16 Chen, Jeng-Shyong, 1995, "Computer-aided accuracy Enhancement for Multi-Axis CNC Machine Tool, *Int J Mach Tools Manufact.,*

17 Buckner, G. D., 2000, "Intelligent Sliding Mode Control of Turning Operations", *Proc ISCA 2000*

18 Demuth, H and Beale, M., Neural Network Toolbox for Use with MATLAB. The MATHWORKS Inc., 1998

19 Mehdi, Q, Dixon, W., Gough, N.E. and . Pitchford, J., 2000a, "Simulating Thermally Induced Error in CNC Machine Tools", **"**, *Proc. European Simulation Symposium and Exhibition: ESS2000,*

20 Bryan, JB. International Status of Thermal Error Research. Annals of CIRP 1968; 16:203-215

21 Dixon W, Mehdi Q, Gough, N, Pitchford J. Adaptive Neuro-Fuzzy Inference Modelling of Errors in CNC Machines. In: Proc ISCA 8th International Conference on Intelligent Systems 1999; pp185-90

22 Weck M, McKeown P, Bonse R, Herbst U. Reduction and Compensation of Thermal Errors in Machine Tools. Annals of the CIRP 1995; 44:part 2 589-598

23 Microsoft 1998 MSDN Library Visual Studio 6 Visual Basic Documentation

24 The Mathworks , 1998, Matlab Application Program Interface Guide Version 5

SESSION 2

MEDICAL, BUSINESS AND PERSONAL APPLICATIONS

A data mining approach to the prediction of corporate failure

Feng Yu Lin and Sally McClean
Faculty of Informatics, University of Ulster, Coleraine, BT52 1SA, UK
fy.lin@ulst.ac.uk or t55294@hotmail.com
si.mcclean@ulst.ac.uk

Abstract: This paper uses a data mining approach to the prediction of corporate failure. Corporate data consisting of failed and nonfailed quoted companies listed on the London Stock Exchange for the latest 20 years (1980 – 1999) was used for the empirical study. This paper initially uses four single classifiers - Discriminant Analysis, Logistic Regression, Neural Networks and C5.0 - each based on two feature selection methods for predicting corporate failure. Of the two feature selection methods - human judgement based on financial theory and ANOVA statistical method - we found the ANOVA method performs better than the human judgement method in all classifiers except discriminant analysis. Among the individual classifiers, decision trees and neural networks were found to provide the better results. Finally, a hybrid method that combines the best features of several classification models is developed to increase the prediction performance. The empirical tests show that such a hybrid method produces higher prediction accuracy than individual classifiers.

1. Introduction

Due to recent changes in the world economy and as more firms, large or small, seem to fail now more than ever corporate failure prediction is of increasing importance. Corporate failure prediction is not only an interesting but also a challenging problem that has led to several studies over the past four decades. Numerous statistical classifiers have been constructed for the prediction of corporate failure. The main techniques used include discriminant analysis [1, 28], logistic regression [19, 23, 32], and probit analysis [9, 27].

Altman [1], Argenti [3] and Lincoln [22] argue that corporate failure is not an instantaneous occurrence but that it is a process which evolves over a considerable period of time. Companies do not fail overnight. Since corporate failure evolves over a considerable period of time, this gives us the foundation for predicting corporate failure. According to UK 1985 Act (section 228): "every balance sheet and profit and loss account shall give a true and fair view", so the financial statement data are a set of facts which should include extensive warning signals pointing toward failures. Using these facts we can extract the discovery of patterns or rules describing a firm's financial status; such patterns or rules potentially leads to decision support using methods such as classification and forecasting.

2. Data mining

Data mining is a set of techniques carried out in a logical order. There are a number of basic tasks that are frequently carried out as part of data mining, many of which may be tackled using a number of different approaches. In this study we follow a procedure for Data Mining: the SAS SEMMA methodology [16]. Figure 1 shows the SEMMA process. This acronym SEMMA is based on the five stages of the data mining process – Sampling, Exploration, Manipulation, Modeling and Assessment. Not all steps are critical to the whole data mining process, however, a data mining project should at least consider all of them.

2.1 Sampling

The data sample consists of company financial data from the UK. The companies are divided into two groups; one is the failed companies group the other is the nonfailed companies group. The nonfailed companies are those listed on the London Stock Exchange, while the failed firms in this study are of those one year prior to failure. They are all UK listed companies of negligible value, or Companies in Receivership and/or Liquidation.

The data collected relates to 1133 companies over the span of the last 20 years, 1980 - 1999. No attempt is made to match the failed and nonfailed companies. Ohlson 1980 [24] states "the appropriate criteria to be used for matching purposes are not obvious". Frydman Alttman and Kao 1985 [14], Taffler 1982 [28], Watson 1995 [29], and Lacher et al. 1995 [20] have previously undertaken this kind of unmatched sample study.

The financial data was accessed from Datastream/ICV, which is part of Primark global information services Corporation (NYSE/PSE: PMK). The attributes used are account items from both the Balance Sheet and P/L (Profit and Loss) Account (i.e. Income Statement). In this study the initial collected variables covering the important areas of overall business performance are listed as in Table 1.

Table 1 Initial Collected Variable Sets

Concept	Variables	Meaning
Rates of Return	X701	RETURN ON S'HOLDERS EQUITY %
	X703	RETURN ON S'HOLDERS CAPITAL %
	X705	RETURN ON LONG TERM CAPITAL %
	X707	RETURN ON CAPITAL EMPLOYED %
	X709	RETURN ON NET FIXED ASSETS %
	X711	TRADING PROFIT MARGIN %
Profit margins	X713	OPERATING PROFIT MARGIN %
	X716	PRE-TAX PROFIT MARGIN %
	X717	NET PROFIT MARGIN %
	X718	EARNINGS MARGIN %
	X719	CASH FLOW MARGIN %
Turnover ratios	X721	TURNOVER/ASSETS EMPLOYED
	X722	TURNOVER/FIXED ASSETS
	X723	TURNOVER/N. CURRENT ASSETS
	X724	STOCK TURNOVER
	X725	STOCK RATIO (DAYS)
	X726	DEBTORS TURNOVER
	X727	DEBTORS RATIO (DAYS)
	X728	CREDITORS TURNOVER
	X729	CREDITORS RATIO (DAYS)
Gearing ratios	X731	CAPITAL GEARING %
	X732	INCOME GEARING %
	X733	BORROWING RATIO
	X734	INCOME GEARING % (INC ASS)
	X735	GCF/TOTAL LIABILITIES
	X736	PREF AND LOAN/EQU.& RESERVES
	X737	LOAN CAP./EQUITY & RESERVES
Liquidity ratios	X741	WORKING CAPITAL RATIO
	X742	QUICK ASSETS RATIO
	X743	CASH/CURR. LIABILITIES
Productivity ratios	X761	TAX RATIO
	X762	SALES PER EMPLOYEE
	X763	OPERATING PROFIT PER EMPLOYEE
	X764	CAPITAL EMPLOYED /EMPLOYEE
	X765	STOCK AND W.I.P. PER EMPLOYEE
	X766	AVERAGE SALARY PER EMPLOYEE
Per-share items and yields	X792	CASH EARNINGS PER SHARE

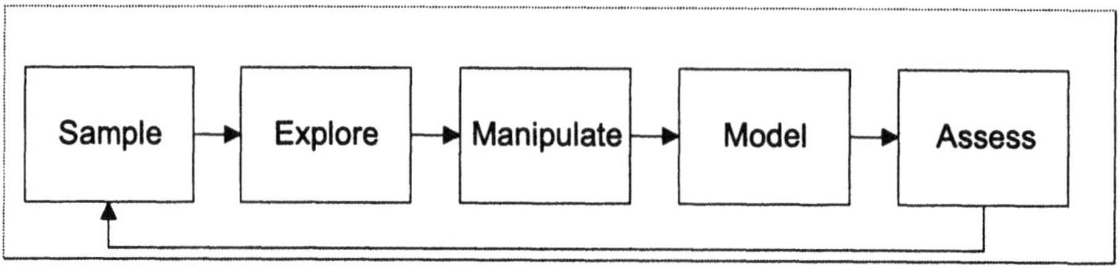

Figure 1 The SEMMA Data Mining Process

This training sample consists of 690 nonfailed companies and 106 failed companies. The data for the failed group relates to *one year before failure*. All the companies' data in the training sample are between 1980 and 1990. Estimated models based on this training sample are used to predict the unseen test dataset which consists of companies' data between 1991 and 1999. Details of the training sample and test sample are shown in Table 2.

Table 2 Training set and Test set

Data sample	No of nonfailed firms	No of failed firms	Time span
Training set	690	106	1980 – 1990
Test set	289	48	1991 – 1999

2.2 Data Exploration

Exploration or preprocessing of data is very important and is sometimes the most time consuming part of the data mining process. Firstly, the data collected is not often presented in a ready to use state, so we need to transform it into a table or database. Secondly, usually the sample data collected contains missing data, this is especially true of data from the financial statement. Missing data is normally expressed as a blank or N/A (Not Available). Some software will treat a blank or N/A as zero, and some will not accept blank or N/A in the numerical field. Hence we need to transform the blank or N/A into a reserved number, e. g. –999.99, which is a surrogate for the missing value. If processed in SPSS, we can put a statement "MISSING VALUE ALL (-999.99)." to specify that the value –999.99 represents a missing value. Thirdly, we must filter out the redundant records such as duplicated data. To filter out such records, we can make use of a unique index, for example, in dBase we can use the command "INDEX ON *fieldname* TO *indexfile* UNIQ" to filter out duplicated data. Fourthly, we need to clean the missing values. In data mining, it is often desirable to eliminate these from the datafile. For eliminating missing values, different techniques are appropriate in different situations such as the datafile size, or the number of fields containing missing values. There are four ways to deal with the missing values:

- Omit the records containing missing values
- Omit the fields containing missing values
- Fill in the missing values with default values
- Fill in the missing values with a value derived from a model.

We use ANOVA (Analysis of Variance) for data exploration. **Table 3** is the ANOVA output for the training sample. This table contains a large amount of information. Firstly, from this table we can understand the descriptive statistics about our training sample set, for example, the mean, standard deviation, and number of cases in the failed and nonfailed group for each independent variable. Secondly, we can ascertain which input variables (financial ratios) are significantly different between the nonfailed and the failed group. Thirdly, we can obtain the information about how many missing values there are in each of the input variables. From **Table 3** we decide to delete X719, X734, X735, X761, X766, and X792 from the training set, since these variables include too many missing values as compared with other variables. The number of missing values in these 6 variables is 369, 186, 369, 56, 151 and 369 respectively.

Table 3 Anova Output of Training Set

X	MEAN NFail	STD Nfail	N NFail	MEAN (Fail)	STD (Fail)	N (Fail)	F	DF	P
X701	18.18	64.59	690	-7.4	126.84	106	10.4581	794	0.0013*
X703	15.71	63.46	690	-9.97	86.31	104	13.3291	792	0.0003*
X705	18.56	59.5	690	3.02	26.9	106	7.005	794	0.0083*
X707	17.81	58.68	690	2.94	26.57	106	6.5991	794	0.0104*
X709	67.95	761.85	690	-16.48	96.79	106	1.2973	794	0.255
X711	0.66	204.36	690	-3.01	65.02	106	0.0335	794	0.8548
X713	-4.33	225.58	690	-6.9	70.49	106	0.0136	794	0.9072
X716	-2.72	193.8	690	-10.03	68.12	106	0.1479	794	0.7007
X717	-6.06	193.49	690	-10.85	67.82	106	0.0639	794	0.8005
X718	-5.71	193.52	690	-10.84	67.85	106	0.073	794	0.7871
X719	-7.09	228.88	389	-14.34	102.0	38	0.0374	425	0.8468
X721	2.57	3.6	690	2.55	2.79	106	0.0023	794	0.9615
X722	8.49	19.55	690	6.07	7.48	106	1.5901	794	0.2077
X723	14.43	155.46	690	-6.85	45.56	106	1.959	794	0.162
X724	21.47	106.63	690	16.78	43.97	106	0.2003	794	0.6546
X725	70.19	98.88	690	81.89	78.12	106	1.353	794	0.2451
X726	7.4	10.26	690	5.82	4.23	106	2.4448	794	0.1183
X727	96.71	437.66	690	96.89	98.07	106	0.0	794	0.9965
X728	5.45	2.83	690	4.45	2.17	106	12.2157	794	0.0005*
X729	90.42	133.78	690	112.98	121.49	106	2.6746	794	0.1024
X731	22.18	107.6	690	61.71	39.36	106	14.0048	794	0.0002*

X	MEAN NFail	STD Nfail	N NFail	MEAN (Fail)	STD (Fail)	N (Fail)	F	DF	P
X732	10.21	269.98	690	52.98	208.18	106	2.4365	794	0.1189
X733	0.49	1.69	690	1.81	8.36	106	13.483	794	0.0003*
X734	5.68	5.43	5	42.65	27.01	5	9.0059	8	0.017*
X735	0.11	0.11	389	0.01	0.2	38	21.3075	425	0.0*
X736	0.27	0.92	690	1.22	7.36	106	10.3693	794	0.0013*
X737	0.21	0.65	690	1.21	6.9	106	13.9304	794	0.0002*
X741	1.62	0.84	690	1.11	0.61	106	35.7554	794	0.0*
X742	1.0	0.72	690	0.65	0.6	106	22.6307	794	0.0*
X743	0.32	0.63	690	0.14	0.52	106	8.2717	794	0.0041*
X761	26.92	29.19	619	17.2	32.5	88	8.2951	705	0.0041*
X762	90.1	287.81	690	65.8	63.97	97	0.6858	785	0.4079
X763	10.5	88.89	690	0.26	16.28	97	1.28	785	0.2582
X764	64.19	339.99	690	53.76	77.81	97	0.0906	785	0.7636
X765	15.3	41.04	690	16.16	32.4	97	0.0392	785	0.843
X766	11.3	7.32	561	11.01	5.23	84	0.1266	643	0.7221
X792	22.08	69.11	389	14.33	37.35	38	0.4638	425	0.4962

2.3 Data Manipulation

One main purpose of data manipulation is to select subsets of attributes of interest in order to reduce the number of predictors to be used in modeling. This variable selection process is called feature selection. Feature selection has long been an active research topic in statistics and pattern recognition [3,17,12]. Feature selection is important in that it can simplify the data description and this in turn makes for easier understanding of problems, and better and faster problem solving. The other reason for reducing the number of independent variables is to prevent multicollinearity between the variables. Here we will use two feature selection methods. The first one is based on financial theory and human judgement, the second is based on ANOVA (analysis of variance analysis).

2.3.1 Feature selection I - human judgement based on financial theory

Traditionally, the independent variables (features) in corporate failure models have been selected empirically in a manner because they were found to be significant in other studies [1, 24]. Laurent 1979 [21], Ezzamel et al 1987 [13] and Clarke 1990 [10] attempted to reduce the wide variety of financial ratios into several major groups (e.g. profitability, liquidity ..., etc.), the suggestion being that a researcher need only select one ratio from each of the groups to obtain an indication of the companies overall performance. There is however no guide as to which ratio should be selected within each group [10]. Therefore, the second feature selection is based on the variables used in financial failure models under the headings of (1) profitability, (2) efficiency, (3) gearing (leverage), and (4) liquidity. The features selected are from these categories respectively, that is

- Return on Capital Employed (ROCE)
- Turnover to Total Assets Employed (Turnover/TA)
- Capital Gearing (Leverage) (CG)
- Working Capital ratio (WC)

2.3.2 Feature selection II - ANOVA statistical method

Here we use ANOVA (analysis of variance analysis) to select the variables that are significantly different between these two groups. To classify failed companies and the nonfailed companies effectively, the predictors should be chosen to enable us to distinguish between these two groups. This is just like two football teams. If one team wears black, the other wears white, then the attribute – the clothes colour - can be used to classify these two teams. This is because the clothes colour is significantly different between these two groups. So, as far as is possible, the variables used in the prediction model should be significantly different in distinguishing between failed and nonfailed groups. From **Table 3**, the ANOVA output of the training set, we can get the variables indicated by * which indicate a significantly different between the failed and nonfailed group.

2.4 Modeling and Results

Although there are many approaches to the classification problem, they can be grouped into two paradigms:

- statistical approaches and mathematical processes,
- machine learning approaches such as rule based methods, and the neural network approach.

So the classification models used in this study are:

Statistics: discriminant analysis, logistic regression.

Machine learning methods: decision trees, neural networks.

2.4.1 Discriminant analysis

In this study, discriminant analysis (DA) is employed to classify firms into one of two groups – failed or nonfailed – based on their financial characteristics. The result of discriminant analysis is shown in **Table 4**. For DA1 (Discriminant Analysis using feature selection 1), the number of correct predictions is 265, but the number of wrong predictions is 72, so 78.6% of the test sample are correctly predicted. For DA2 (Discriminant Analysis using feature selection 2), The number of correct predictions is 261, whereas the number of wrong predictions is 76, and so 77.4% of the test sample is correctly predicted.

Table 4 Prediction of single classifier based unseen test sample

Model	No of correct predictions	No of wrong predictions	% of correct predictions	% of wrong predictions
DA1	265	72	78.6	21.4
DA2	261	76	77.4	22.6
LG1	283	54	84.0	16.0
LG2	285	52	84.6	15.4
NN1	295	42	87.5	12.5
NN2	297	40	88.1	11.9
C51	292	45	86.6	13.4
C52	299	38	88.7	11.3

DA1, LG1, NN1, C51 are feature selection by human judgement,
DA2, LG2, NN2, C52 are feature selection by ANOVA
DA (discriminant analysis), LG (logistic regression), NN(neural network), C5 (decision tree)

2.4.2 The Logistic regression model

The logistic regression or logit model is based on the cumulative logistic probability function. It has been found to be appropriate in many situations involving a binary dependent variable (e.g., failed or nonfailed). The result of logistic regression is shown in **Table 4**. For LG1 (Logistical regression using feature selection 1), the number of correct predictions is 283, whereas that of incorrect predictions is 54, therefore 84.0% of the test sample is correctly predicted. For LG2 (Logistical regression using feature selection 2), the number of correct predictions is 285, but the wrong prediction is 52, so 84.6% of the test sample is correctly predicted.

2.4.3 Neural networks

Neural Networks' strengths and weaknesses accrue from the fact that they need no prior model assumptions, and from their capability to infer complex, nonlinear underlying relationships. The neural networks model has good performance in domains where independent variables (financial indices) and a dependent variable (failed or nonfailed) have non-linear relationships and where the correlation between independent variables are high [26].

The result of neural network analysis is shown in **Table 4**. For NN1 (Neural network using feature selection 1), the number of correct predictions is 295, but the number of wrong predictions is 42, and so 87.5% of the test sample are correctly predicted. For NN2 (Neural network using feature selection 2), the number of correct predictions is 297 and the number of wrong predictions is 40, so 88.1% of test sample is correctly predicted. The result of the neural network is therefore superior to discriminant analysis and the logit model.

2.4.4 Decision trees – C5.0

Decision trees are a rule induction method. They are hierarchical, sequential classification structures that recursively partition the learning sets [8, 25]. The result of using decision trees is shown in **Table 4**. For C51 (Decision tree C5.0 using feature selection 1), the number of correct predictions is 292, that of wrong predictions is 45, and therefore 86.6% of the test sample is correctly predicted. For C52 (Decision tree using feature selection 2), the number of correct predictions is 299, that of wrong predictions is 38, thus 88.7% of the test sample is correctly predicted.

2.4.5 Hybrid Models combining different classifiers

Just like in a management meeting, the best decision may not be the individual's opinion but the group aggregated opinion. A simple way to take account of different opinions is to take the vote of the majority of the population of classifiers or of a subset of it. In this study, we will employ voting with different weights, using the hybrid classifier to predict the unseen test data.

Let us suppose we have m classifiers and V_{ij} is the predicted output that classifier or method j predicts on the ith record or company in the unseen test sample. O_i is the predicted output that the hybrid classifier or method makes on the ith record in the unseen test sample. We will suppose the cost of Type I and Type II errors to be equal. If our purpose is to minimize the number of wrong predictions, then the problem can be formulated as following:

$$\text{Min} \qquad \delta = \sum_{i=1}^{n} T_i \oplus O_i$$

s.t.

$$O_i = f(w_1V_{i1}+w_2V_{i2}+w_3V_{i3}+ \ldots\ldots +w_mV_{im} - \theta)$$

Where

$\oplus$ = Is the exclusive OR function. this function $\oplus$ returns 1 if both side of $\oplus$ is different, returns 0 if both side of $\oplus$ is the same.

f = Is the Heaviside funcion to give failed or nonfailed index

θ = Threshhold or critical value

δ = Miss hit (misclassification) of hybrid classifier

T_i = 0 or 1 - acutual outcome of ith company in the test sample

O_i = 0 or 1 - predicted output of ith company in the test sample based on the hybrid method.

$O_i = 1$ If $w_1V_{i1}+w_2V_{i2}+w_3V_{i3}+ \ldots\ldots +w_mV_{im} > \theta$, $O_i = 0$ else

V_{ij} = Predicted output of ith company in the test sample by classifier j

w_j = Weight of classifier j

m = Number of independent classifiers

n = Number of records or companies in test sample

The problem of finding such a combination of classifiers is to try to find (1) the combination of weights (2) their respective classifiers (3) the threshhold θ such that the miss hit δ between actual output T_i and O_i is as small as possible. The whole process is very similar to the discrimination analysis or logistic model. Here the weight we adopt is simply the prediction accuracy of an individual's classifier in the training sample. Since the test sample is unseen data before we do the prediction, the information we have is performance in the training phase. The better the performance, the more weight in the voting. The training accuracy of the training samples is shown in Table 5. In conclusion the proposed method can be described algorithmically as follows:

Table 5 Training accuracy of training set

Classifier	Human Feature selection 1	ANOVA Feature selection 2
DA	81.0	84.9
Logistic	86.2	87.8
Neural networks	87.7	89.6
Decision trees	90.1	89.6

The Hybrid Algorithm

(1) Compute the total hit ratio (total accuracy) for the training sample itself for all m independent classifiers.

(2) Take the outcome prediction V_{ij} of classifier j for all the records i in the unseen test sample for $j = 1, .. m$, $i = 1, .. n$.

(3) Take all the population of classifiers, or subsets of it. Compute $w_1V_{i1}+......+w_jV_{ij}+ ....+w_mV_{im}$ for each record i in the test sample. Take the total accuracy for classifier j obtained from (1) as w_j.

(4) If $w_1V_{i1}+w_2V_{i2}+w_3V_{i3}+ ....+w_mV_{im} > \theta$, then Oi=1, else Oi = 0. Adjust the parameter θ, where $0 < \theta < w_1+w_2+w_3+ ....+w_m$, such that the miss hit (misclassification) δ is as small as possible.

Here we present three hybrid classifiers. They are the combination or linear transformation of single classifiers, the weight is their prediction accuracies in the training sample.

Hybrid 1	2 statistics+2 machine learning (all):	DA + Logit + NN + C5.0
Hybrid 2	1 statistics+2 machine learning:	DA + NN + C5.0

Hybrid 3 1 statistics+1 machine learning: Logit + C5.0

Table 6 Prediction based on hybrid method

Feature Selection	Classifier	Hybrid model	No of correct predictions	No of wrong predictions	% of correct predictions
1	Hybrid 1	$a_1DA+a_2LG+a_3NN+a_4C5$	298	39	88.4
1	Hybrid 2	$a_1DA+a_3NN+a_4C5$	297	40	88.1
1	Hybrid 3	a_2LG+a_4C5	297	40	88.1
2	Hybrid 1	$b_1DA+b_2LG+b_3NN+b_4C5$	302	35	89.6
2	Hybrid 2	$b_1DA+ b_3NN+b_4C5$	302	35	89.6
2	Hybrid 3	b_2LG+b_4C5	301	36	89.3

a_1 a_2 a_3 a_4 are training accuracy of feature selection 1 for DA, Logit, NN, C5.0 respectively.
b_1 b_2 b_3 b_4 are training accuracy of feature selection 2 for DA, Logit, NN, C5.0 respectively. Ref Table 5

The results of the hybrid method are shown in Table 6. For feature selection 1 (financial and human), the prediction accuracy for hybrid 1, hybrid 2 and hybrid 3 are 88.4%, 88.1%, 88.1% respectively for the test dataset. They are better than the best performance 87.5% for neural networks. For feature selection 2 (ANOVA), the prediction accuracy for hybrid 1, hybrid 2 and hybrid 3 are 89.6%, 89.6%, 89.3% respectively. They are also better than the best performance 88.7% for single classifier C5.0.

2.5 Assessment

In this study, we applied four classifiers - discriminant analysis (DA), logistic regression (logit), neural networks (NN) and decision trees C5.0 for predicting corporate failure. Among these four classifiers, we find that machine learning methods – C5.0 and neural networks - outperform discriminant analysis and logistic regression. As shown in Table 7, for the feature selection 1, the prediction accuracy of DA, logit, NN and C5.0 are 78.6%, 84%, 87.5% and 86.6% respectively. For the feature selection 2, the prediction accuracy for DA, logit, NN and C5.0 are 77.4%, 84.6%, 88.1% and 88.7% respectively. But the accuracy gained by machine learning methods is a time consuming process through adjusting the model parameters – by trial and error. The human time and labour spent on these two methods are greater than for statistical methods. The three hybrid classifiers in this study are based on the same algorithm but with different combinations and parameters. Accordingly, in Table 7, for feature selection 1, the prediction accuracy of hybrid 1 classifier to hybrid 3 classifier is 88.4%, 88.1% and 88.1% which are better than single classifiers DA (78.6%), Logit (84.0%), NN (87.5%) and C5.0 (86.6%). For feature selection 2, the prediction accuracy of classifiers hybrid 1 to hybrid 3 is 89.6%, 89.6% and 89.3% which are also better than single classifiers DA (77.4%), logit (84.6%), NN (88.1%) and C5.0 (88.7%).

Table 7 Comparison of Prediction Accuracy

Method	Human Feature Selection 1	ANOVA Feature Selection 2
DA	78.6	77.4
Logistic	84.0	84.6
NN	87.5	88.1
Decision tree	86.6	88.7
Hybrid 1	88.4	89.6
Hybrid 2	88.1	89.3
Hybrid 3	88.1	89.6

3. Discussion and Conclusion

In this study we use two feature selection method to choose independent input variables. The first one is human judgement based on financial theory, which has been used by many researchers before, the second one is the ANOVA statistical method. For all the models – DA, LG, NN, C5.0 and hybrid classifiers, we found the ANOVA feature selection is better than human judgement feature selection except for DA (Discriminant Analysis). The reason may be up to now there is no complete integrated theory of corporate failure that exists. So using the ANOVA computing method to select features gives a better performance.

Also, we analyzed the performance of several models developed for the problem of corporate failure prediction. The machine learning methods (neural networks and decision trees) show better performance than the statistical approach. We have suggested the hybrid algorithm that combines statistical and machine learning methods. In this algorithm, discriminant analysis, logistic regression, neural network, and decision tree C5.0 can be used as the base classifier to produce a hybrid classifier. The hybrid 1 classifier is the combination of four classifiers: DA, Logit, NN, and C5.0. The hybrid 2 classifier is the combination of 3 classifiers: DA, NN, and C5.0 and the hybrid 3 classifier is the combination of 2 classifiers – Logit and C5.0. We used real world data as training samples plus these three different hybrid classifiers to show its effectiveness. The main finding is that the hybrid methods produce better results in the prediction of corporate failure at one year before bankruptcy.

Reference:

[1] Altman, EI. 1968. *Financial Ratios, Discriminant Analysis and the Prediction of Corporate Bankruptcy*, The Journal of Finance, Vol. XXIII, Sep 1968, No4, pp589-609

[2] Altman EI. 1980. *Commercial Bank Lending:Process, Credit Scoring, and Costs of Errors in Lending*. Journal of Financial and Quantitative Analysis 15 (November): 813-832.

[3] Andrew Webb, Statistical pattern recognition, Arnold London, co published by Oxford Univ Press Inc. New York, 1999 pp 213 -269.

[4] Argenti, J. 1976. Corporate Collapse: The Causes *and Symptoms*, New York: John Wiley & Sons.

[5] Ball, R. and G. Foster. 1982. *Corporate Financial Reporting: A methodological Review of Empirical Research*. Journal of Accounting Research, Vol. 20 (Supplement): 161 - 234.

[6] Bierman H, and Hausman WH. 1970. *The credit granting decision*. Management Science, 16: 519-532

[7] Braun, H. and Chandler, J.S. 1987. *Predicting Stock Market Behavior through Rule Induction: An Application of the Learning-From-Example Approach*. Decision Science 18: 415-429.

[8] Breiman JR, Friedman JH, Olshen RA, Stone Ch. J. 1984. *Classification and Regression Tress*, Chapman & Hall, Inc., New York 1984.

[9] Casey M, McGee V and Stinkey C. 1986. *Discriminating between Reorganized and Liquidated Firms in Bankruptcy*. Accounting Review, April: 249-262.

[10] Clarke, PJ. 1990. *An Empirical Analysis of Useful Financial Ratios: an Exploratory Study of Irish Public Quoted Companies*. Irish business and administrative research, vol 11, 40-57.

[11] Crasic, M. R. and Perrault, W. D. 1977. *Validation of Discriminant Analysis in Marketing Research*, Journal of Marketing Research, 14: 60-68.

[12] Devijver PA and Kittler J. 1982. *Pattern Recognition: A Statistical Approach*. Prentice Hall International.

[13] Ezzamuel M, Brodie J, and Mar-Monlinero C. 1987. Financial patterns of manufacturing companies. Journal of Business Finance and Accounting. pp 519-536.

[14] Frydman H, Altman EI, and Kao D. 1985. *Introducing Recursive Partition for Financial Classification: The Case of Financial Distress*. the Journal of Finance, vol. XL, No. 1, March: 269-291.

[15] Garrison L, Michaelson R. 1989. *Symbolic Concept Acquisition: A New Approach to Determining Underlying Tax Law Construct*. Journal of American taxation association. Fall: 77-91.

[16] Held G. 1997. *The Process of Data Mining*, VIII international Symposium on Applied Stochastic Model and Data Analysis: 155 – 164.

[17] Jain A and Zongker D. 1997. Feature Selection: Evaluation, Application, and Small Sample Performance, IEEE Trans. on Pattern Analysis and Machine intelligence, vol. 19, no. 2: 153-158.

[18] Karels GV and Prakash A.1987. *Multivariate Normality and Forecasting of Business Bankruptcy*, Journal of Business Finance and Accounting, winter.

[19] Keasey K, McGuinness P and Short H. 1990. *Multilogit Approach to Predicting Corporate Failure – Furthur Analysis and The Issue of Signal Consistency.* Omega 18(1): 85-94.

[20] Lacher R.C., Pamela K. Coats, Shanker C. Sharma and L. Franklin Fant.1995. *A Neural Network for Claassifying the Financial Health of a Firm.* European Journal of Operational Research 85: 53-63.

[21] Laurent, CR. 1979. Improving the Efficiency and the Effectiveness of Financial Ratio Analysis. Journal of Business Finance and Accounting, 401-413.

[22] Lincoln,M. 1984. *An Empirical Study of the Usefulness of Accounting Ratios to Describe Levels of Insolvency Risk*, Journal of Banking and Finance, 1984:321-340.

[23] Martin D. 1977. *Early Warning of Bank Failure: A Logit Regression Approach.* Journal of Banking and Finance 1: 249-276.

[24] Ohlson JA. 1980. *Financial Ratios and The Probabilistic Prediction of Bankruptcy.* Journal of Accounting Research, Spring: 109-131.

[25] Quinlan JR. 1993. *C4.5 Programs for Machine Learning*, Morgan, California.

[26] Salchenberger, L. M., Cinar, E. M., and Lash, N. A., Neural Networks: A New Tool For Predicting Thrift Failures, Decision Sciences, 23, 1992, 899-915.

[27] Skogsvik K. 1990. *Current Cost Accounting Ratios as Predictors of Business Failure: The Swedish Case.* Journal of Business Finance and Accounting 17(1): 137-160.

[28] Taffler RJ. 1982. *Forecasting Company Failure in the UK using Discriminant Analysis and Financial Data.* Journal of Royal Statistics Soc. A, 124, Part 3: 342-358.

[29] Watson ID. 1995. *An Investigation of The Use of Market and Industry Data in Financial Distress Modeling: Based on Data Derived from The Unlisted Securities Market and Official List*, Ph.D. dissertation, school of management, Univ. of Ulster.

[30] Webb A. 1999. *Statistical Pattern Recognition*, Arnold London, co published by Oxford Univ Press Inc. New York: 213 -269.

[31] Welker RB. 1974. *Discrimination and Classification Analysis as an Aid to Employee Selection*, Accounting Review, 49: 514-523.

[32] Zavgren CV. 1985. *Assessing the Vulnerability to Failure of American Industrial Firms: A Logistic Analysis.* Journal of Business Finance and Accounting 12(1): 19-45.

Resolving Deontic Operator Conflicts in Legal Documents

J.T. Sykes & V. Konstantinou

Harrow School of Computer Science
University of Westminster
Northwick Park, Watford Road, Harrow
Middlesex HA1 3TP,England
sykesj@wmin.ac.uk & vassilis@wmin.ac.uk
www.Wmin.ac.uk/CS

Abstract: This paper presents experimental work on the development of a system that is aimed at (semi)-automatically identifying deontic inconsistencies in legal documents. The target text with which the system was tried was the British data protection Act (1984).

The motivation behind our project was to assist authors of such documents to identify and hopefully resolve inconsistencies and ambiguities before the document in question is released to the target audience.

The main difficulty in writing legal documents does not lie in the structural complexities of the language, but rather in the meaning of the phrases used. A particular problem arises in the form of deontic operators, which may mislead the unwary reader especially if he/she is not familiar with the legal domain.

Our system attempts to solve potential problems of this kind, using both the syntactic and semantic information contained within a natural language legal text. The system employs the Conceptual Graph knowledge representation formalism in conjunction with a modified parser to create structures, which can subsequently be compared at the semantic level, regardless of differences in the surface syntax.

1. Introduction

The increasing desire for legal documents to be couched in language that is accessible to the non-legal specialist brings not only benefits, but also certain difficulties to the interpretation of the information concerned therein. The use of shorter, more fragmented and less formal documents allows for the introduction of not only ambiguity, but also of contradiction, or at best inconsistency of guidance.

Whilst this may not be a prevalent problem for those involved in jurisprudence, it can be a major problem for those whose duty it is to follow those guidelines/regulations. A prime example here may be the British Data Protection Act (1984), which, in many cases is applicable to individuals and small companies who may not be fully versant with the legal language. In this particular case, the laudable efforts to make the regulations readable by such laypersons, has the potential of introducing inconsistent information that may well result in an incorrect interpretation of the need to register under the Act.

Whilst many of these problems may be identified by careful proofreading or indeed by the spell-checkers and thesauruses generally available in word processors, the true difficulty does not lie in such *syntactic* (structural) distinctions, but rather in the *semantics* (meaning) of the phrases used. Even the best of proof-readers are apt to miss some of the subtleties involved in the eight guidelines of the British Data Protection Act – and no spell-checker will pick up deontic (strength of obligation, in this context) inconsistencies between paragraphs.

For this reason, we have developed a system (written in Prolog) which is aimed at (semi)-automatically identifying potential problems of this kind, using both the syntactic and semantic information contained within a natural language legal text. The system employs the Conceptual Graph knowledge representation formalism developed by John Sowa [25] in conjunction with a modified parser to create structures, which can subsequently be compared at the semantic level, regardless of differences in the surface syntax.

It is important to note that, although the final interpretation of the results will of course rest with a human expert, there is no need to change the original documentation – by employing mark-up languages, for example - for this system to identify *potential* problems with documentation. Nor is there any suggestion that this system can fully automate the process: The legal expert will always need to have the last say in the interpretation of the results.

2. The British Data Protection Act

The British Data Protection Act (1984)[1] was selected for two reasons. It is a complex piece of legal writing, being composed of eight distinct sets of (cross-referencing) guidelines. It is suggested that documents in this form would be the

---

[1] The 1985 act has now been replaced by the 1998 act. The work described in this paper was based on the text of the original act.

ideal target for a system to check for inconsistencies, since it is notoriously difficult to maintain consistent language – especially as multiple authors may be involved - and to proof read over multiple documents.

Also, the British Data Protection Act was specifically written to be of a readable form, and to avoid too much legal jargon. In doing this, the authors faced the difficulties of non-formalised language, and as has been suggested, this is where the inconsistencies may arise more readily. Although inconsistencies may of course be found in formal language, or indeed be deliberately introduced, there is more scope for the problem to arise, and be missed by human proofreaders in less formalised systems.

In looking at the British Data Protection Act, it was indeed found that there is great potential for such deontic inconsistencies in multiple document guidelines or regulations. The following example occurs early in the act, where much use of the very strong operator "must" is made:

> *"Every data user who holds personal data must be registered, unless all the data are exempt" [British Data Protection Act 1984, Guideline 1, Second Series, Feb. 1989, Section 3].*

Figure 1: Strong Deontic Operator

the weaker "should" is introduced – for example when talking about exemptions for unincorporated members' clubs:

However, in later guidelines, less use of "must" occurs, and

> *"...If any member does object then either*
>
> *The personal data relating to that member should be taken off the computer...." [British Data Protection Act 1984, Guideline 6, Second Series, Feb. 1989, Section A.6.2].*

Figure 2: Weaker Deontic Operator

In the case of Figure 1, the reader may be sure of their obligation to register the data, but in the case of Figure 2, they may be left in some doubt as to just how much obligation is s/he under to actually remove this information. Is it a legal requirement, or not?

Even more of a potential problem arises when very similar statements arise in different parts of a document. In Guideline 3 of the Act, for example, the following statement is found:

> *"Every data user who holds personal data must, unless all the data are exempt, apply for registration"* [British Data Protection Act 1984, Guideline, second Series, Feb. 1989, Section 1.2].

Figure 3: Alternate Syntax to Figure 1

In this case, the operators are consistent, although the syntax has changed, and our system would let this pass. However, consider the scenario where this statement had been written with less regard to the strength of the operators, and the problem becomes obvious:

> *"Every data user who holds personal data should, unless all the data are exempt, apply for registration"*

Figure 4: Hypothetical Deontic Problem

Here, the confusion to the reader could be potentially be mislead to believe that it is not obligatory to register under the act at all.

It may be argued that the case illustrated in Figure 1 and Figure 4 is a relatively easy problem to spot if the reader is aware of it. However, other deontic terms may be couched in the form of phrases, rather than single words, such as "are obliged to", "ought to", "are required to". Once again, the phrase taken as a whole has a degree of strength that must be kept consistent throughout the legal document(s), if the reader is to be expected to comply with them.

Of course, here, common sense suggests that the "must" has priority over the "should", although the two statements appear to be saying the same thing.

However, these sentences may appear in two different sections of the act, indeed, in two different Guidelines. Details such as these can lead to confusion, at best, and at worst, to loopholes in the law – leading to long and costly debating.

It is apparent from looking at large legal documents, that there is scope for a number of such small, but nonetheless important inconsistencies of language to creep in, which are extremely difficult to spot at the proof reading stage.

Since these inconsistencies tend to be at the semantic level, the Conceptual Graph representation; a formalism devised to represent the relationships between *concepts*, rather than words, offers an ideal framework around which to build a tool to identify these problems, which are difficult for a human proofreader to identify.

Deontic terms by definition have degrees of weight or importance, and in general reading, this may be apparent from the context. However, in the legal domain, such operators take on a somewhat more definite and possibly dangerous role. For example, the operators "must" and "should", in casual writing may well carry very similar semantic force, but in a legal text they may be somewhat different, or indeed the basis for argument and litigation, if not conviction.

For a legal text to state that an individual or vendor *must* perform some action is a binding statement, yet it could be argued that to say that they *should* perform that action is merely advisory in strength.

3. Conceptual Graph Matching for Consistency Checking

It is obvious from the similarity of the statements shown in Figure 1 and Figure 4 that the Conceptual Graph representations should also be very similar. Only the deontic operators are in fact different. Although the structure of the two sentences is slightly different, this dissimilarity is at a purely surface, syntactic level, and the graphs produced by the parser to represent the semantics differ only in the concept representing the deontic operators "must" and "should".

It is apparent, therefore, that the identification of inconsistencies due to the presence of deontic operators can be reduced to a form of graph matching procedure. It is of course acknowledged that multiple differences could occur between any two sentences, due to the presence of multiple deontic operators, but this should also be readily apparent from the Conceptual Graph form. Problems and inconsistencies due to other linguistic factors, such as ambiguity, anaphora and cataphora are somewhat beyond the scope of the investigation documented here, and therefore will not be fully addressed in this paper.

4. Conceptual Graphs

In order to represent the *semantic* information present in natural language legal texts, for manipulation by a computer, it is necessary to employ some form of formal description. This *knowledge representation language* allows for the specification of the data, and thus, the manipulation of the information by automated processes.

Since there is no universal agreement on a psychologically real knowledge representation paradigm, such languages are commonly based on partial theories, and computational models. Traditional representation formalisms include predicate calculus, frames [14], semantic networks [19] and scripts [21].

As each of these methods has its own merits and own inherent problems, there is little agreement as to which may be the most appropriate choice for any given problem. Predicate calculus and its extensions, for example, is a powerful representational technique for a *limited* subset of semantics (the declarative sentence, in its simplest form). It is, however, difficult for the non-specialist to read and to equate back to the original natural language sentence, and has no way of representing contextual dependencies.

As an alternative, the theory behind the knowledge representation formalism of John Sowa: *Conceptual Graphs* [25,27] is mainly concerned with the semantic representation of language, and primarily evolved from semantic networks.

However, Conceptual Graphs have a direct mapping to the predicate calculus, which gives strong theoretical underpinning to the formalism, and they are, in general, considerably less cumbersome and easier to read than equivalent knowledge representation paradigms.

They also, as we will suggest here, enable the comparison of not only single declarative sentences, but also of larger textual structures: paragraphs and anaphoric references through the use of graph matching techniques. The theory generalises on many of the ideas present in preceding work on semantics, such as Fillmore [9], Shank [20] and Montague (see [5]), making it a promising model for representing such information in a computational form.

This is a relatively young, but rapidly expanding representation technique which, due to these advantages, is now being applied in diverse fields, such as Knowledge Acquisition [2,16], Knowledge-Based Systems [24, 23, 11] and most especially in natural language processing [8, 12, 26, 29].

Conceptual Graphs evolved from Pierce's existential graphs (see [25, 27] for details). Although logically equivalent to first order predicate logic, Conceptual Graphs have the advantage of stressing the semantic roles of concepts, through a direct mapping from natural language. There is considerable theory behind the Conceptual Graph representation, in the form of logical, philosophical, psychological and mathematical foundations. The interested reader is referred to [25,27] for more information.

5. Notations for Conceptual Graphs

At the basic level, Conceptual Graphs are simple (though possibly cyclic) directed graphs consisting of concepts linked by binary relations. The graphical representation of Conceptual Graphs can be performed in multiple ways, with the fully graphical representation employing boxes around concepts, and circles enclosing relations.

Figure 5 shows the sentence "A data-user holds data" in this format.

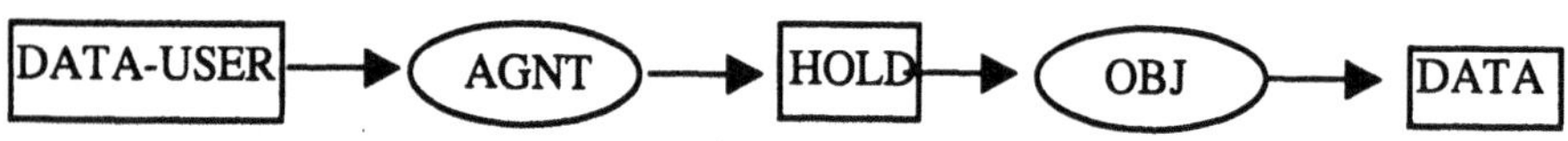

Figure 5: Box and Oval representation of Conceptual Graph

It is important to note that all concepts in this representation are existentially quantified, and thus it is not necessary to specify that a dog or indeed a man exists in order to represent this sentence, in contrast to the traditional Predicate Logic notation.

This is an easy way to interpret graphs, but an alternative convention is often employed, which is more convenient to type on a computer screen, and which translates in a natural manner into list representations, such as used by programming languages such as Prolog. Under this convention, concepts are shown graphically enclosed in square brackets, and relations are enclosed by round brackets (parentheses). Thus, the example of

Figure 5 would be represented in this format as shown in Figure 6.

[DATA-USER] → (AGNT) → [HOLD] → (OBJ) → [DATA]

Figure 6: Bracket representation for Conceptual Graph

6. Ontology and the Type Lattice

Fundamental to the operation of any representation formalism is the ontology behind it. It provides information about the nature of the content and the structure of the information for which the representation provides the form: As Sowa states:

"It is the study of existence, of all the kinds of entities - abstract and concrete - that make up the world" [28, p.51]

In the Conceptual Graph formalism, the ontology is represented in the form of a type lattice.

Every concept in a Conceptual Graph has a corresponding entry in this hierarchical arrangement of types. The entries within the lattice represent the position of a concept relative to others, and maintain the higher order relationships; not between individuals, but between *types* of individuals. This is essential information not simply for interpreting the graphs, but also in order to perform the algorithmic operations that make the system suitable for computation.

Much of the Conceptual Graph literature is concerned to some degree with issues of ontology. For further discussion and theoretic results concerning ontology, type lattices and Conceptual Graphs, the reader is referred especially to [25,27,28(esp. Chapter 2 and Appendix B), 7, 13,31].

7. Conceptual Graphs For Parsing

Conceptual Graphs are particularly suited to the field of natural language processing, due to the close relationship between the orthographic and conceptual representations – a vital relationship in the legal domain: The system does not lose any of the information contained within the original text.

Many systems have been developed which parse natural language into, or generate from, a Conceptual Graph (or partial Conceptual Graph) representation [30, 15, 26]. However, many of these systems require the grammar writer or knowledge engineer to provide quite complex definitional graphs as schemata for each lexical entry.

The system described here allows the concurrent syntactic and semantic parsing of a sentence into a frame of Conceptual Graphs, retaining structural information from a small set of "canonical" (generic) graphs. The use of an intermediate representation removes a level of complexity from the process, and allows a more traditional linguistic approach to grammar development.

The process of extracting a Conceptual Graph (CG) representation for a piece of natural text is performed by two parsing systems, sharing a common lexicon.

Initial syntactic parsing of a sentence is implemented by a Definite Clause Grammar (DCG) [17] in Prolog that produces a structural representation of the syntactic roles of elements of the text.

In view of the fact that the structure of the input can be complex, it was desirable to produce a frame-like representation of the information, holding not only the semantic, but also functional and structural details. The sentence 'every data user who holds personal data must, unless all the data are exempt, apply for registration', can be represented by the phrase-structure rule shown in Figure 7

```
IF {every} data user
      AND holds personal data
      AND NOT  all the data are exempt
THEN must apply for registration.
```

Figure 7:Phrase-structure rule representation

The premise of this investigation is that, if such structures can be created for the set of all meaningful sentences within a legal text, then it is possible to compare, not only the syntactic information they contain but, in conjunction with a powerful semantic representation formalism, the conceptual and intentional force.

The DCG parser is designed to perform the first of these operations; to create a structural representation of the functional roles of clauses, and sentence fragments. These "chunks" of natural language are subsequently further analysed by the Conceptual Graph parser, to provide the semantic information.

The grammar consists of standard DCG rules, building a Prolog structure to represent the parse tree in the Left Hand Side (LHS) arguments, and uses the ability of DCGs to be augmented with Prolog goals to aid functionality. (See [4] for more details).

Since the CG parser must contain a full lexicon in order to parse the text, it would be inefficient, especially in terms of maintenance, consistency and expansion, to provide a separate lexicon for the DCG parser. Thus, the ability to extend the DCG formalism to call external Prolog goals was exploited, to enable the use of the same lexicon by both parsing systems.

The final output from the DCG parser is a Prolog structure, representing the parse tree for the sentence, but also augmented with information about "connective phrases": That is, the sentence segments identified as elements of the phrase-structure rule representation above. An example of a fragment of the representation of the sentence of Figure 3, would be as shown in Figure 8.

```
sent(s(s(np(np(det(every),np(n(data),n(user))),cp(
c(and(who)),vp(v(holds),ap(adj(personal),np(n(data
))))))),cp(c(then(must)),vp(mod(must),vp(v(be),np(n
(registered)))))).
```

Figure 8: Prolog Structure Representation

The resulting structures from this analysis are passed into the second parsing stage. This consists of a nondeterministic LR-1 shift-reduce tabular parser (see [1, 6, 10]) operating within an A/B Unification Categorial Grammar framework [32, 3, 33]. The grammar in its turn is represented in the PATTR-II formalism [22] and modified to produce a Conceptual Graph representation of semantic information, in addition to a full syntactic parse.

The resulting Conceptual Graph representations can then be compared, and any discrepancies found reported to the legal expert.

The overall structure of the system is shown in Figure 9.

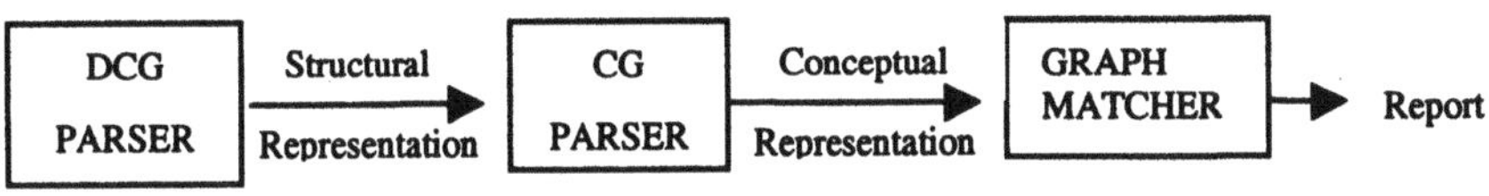

Figure 9: Structure of the CG Parsing System

8. The Conceptual Graph Parser

The mechanism for parsing the sentence segments and producing a Conceptual Graph semantic representation exploits the A/B Categorial Grammar represented in a modified version of the PATTR-II formalism. These modifications allow the representation of the semantics as Conceptual Graphs, and the performance of operations upon these graphs as a part of the parsing process.

The PATTR-II formalism allows the abstraction of the grammar away from the operation of the parser. Thus the author of a grammar need not be conversant with the operation of the parsing process. This allows great flexibility across domains: Since the grammar is not in itself an integral part of the parser, it is possible to substitute other lexicons for different domains, without having to consider the parser itself.

The grammar utilises Head-driven phrase structure grammar (HPSG, [18]) style Attribute-Value Matrices (AVMs). This, being a unification based grammar formalism, is ideally suited to a Prolog implementation. The use of the AVM allows great flexibility, and provides a suitable route for the substitution of Conceptual Graphs for the more traditional relational semantic information.

Figure 10 shows the data structures used by the CG parser for the examples of Figure 3 and Figure 4.

```
[['TEXT',:,[every,data,user,who,holds,personal,data,must,be,registered,',',unless,all,the,data,
are,exempt]],

['GRAPHS',:,

[[[every,data,user],

        [conc('data-user' : [{*},'A',1])]],

[and(who),

[holds,personal,data],

        [conc(personal : [*,5]),rel(attr,[*,4],[*,5] :
        nil),conc(hold : [*,1]),rel(agnt,[*,1],[*,2] :
        nil),conc(animate : [*,2]),rel(obj,[*,1],[*,4] :
        nil),conc(data : [*,4])]],

[then(must),

[must,be,registered],

        [rel(oblig,[],[[conc(be : [*,1]),rel(agnt,[*,1],[*,2] :
        nil),conc(animate : [*,2]),rel(obj,[*,1],[*,4] :
        nil),conc(register : [*,4])],1] : must),conc(act :
        [[conc(be : [*,1]),rel(agnt,[*,1],[*,2] :
        nil),conc(animate : [*,2]),rel(obj,[*,1],[*,4] :
        nil),conc(register :[*,4])],1])]],

[and_not(unless),

[unless,all,the,data,are,exempt],

        [rel(neg,[],[[conc(data:[{*},'A',1]),rel(attr,[{*},'A',
        1],[*,2] : are),conc(exempt : [*,2])],1] :
        unless),conc(entity : [[conc(data :
```

```
        [{*},'A',1]),rel(attr,[{*},'A',1],[*,2] :
        are),conc(exempt : [*,2])],1])]]]]].

============================================================
==================

[['TEXT',:,[every,data,user,who,holds,personal,data,should,be
,registered,',',unless,all,the,data,are,exempt]],

['GRAPHS',:,

[[[every,data,user],

        [conc('data-user' : [{*},'A',1])]],

[and(who),

[holds,personal,data],

        [conc(personal : [*,5]),rel(attr,[*,4],[*,5] :
        nil),conc(hold : [*,1]),rel(agnt,[*,1],[*,2] :
        nil),conc(animate : [*,2]),rel(obj,[*,1],[*,4] :
        nil),conc(data : [*,4])]],

[then(should),

[should,be,registered],

        [rel(oblig,[],[[conc(be : [*,1]),rel(agnt,[*,1],[*,2] :
        nil),conc(animate : [*,2]),rel(obj,[*,1],[*,4] :
        nil),conc(register : [*,4])],1] : must),conc(act :
        [[conc(be : [*,1]),rel(agnt,[*,1],[*,2] :
        nil),conc(animate : [*,2]),rel(obj,[*,1],[*,4] :
        nil),conc(register : [*,4])],1])]],

[and_not(unless),

[unless,all,the,data,are,exempt],

        [rel(neg,[],[[conc(data :
        [{*},'A',1]),rel(attr,[{*},'A',1],[*,2] :
        are),conc(exempt : [*,2])],1] : unless),conc(entity :
        [[conc(data : [{*},'A',1]),rel(attr,[{*},'A',1],[*,2] :
        are),conc(exempt : [*,2])],1])]]]]]
```

Figure 10:Sample CG Parser Data structure for Figure 3 and Figure 4

The current system is capable of automatically producing such structures and identifying deontic conflicts. It does not try to resolve them but to highlight them for the expert user, although in a pretty-printed version of data structure actually shown in Figure 10.

9. Summary and future Work

In the course of the work documented here, a new higher level domain-independent knowledge representation formalism has been employed to provide a very flexible, automated system for the recognition of inconsistencies in natural language texts: Specifically in this current context normative texts, although the system itself is domain independent.

The current system uses the Conceptual Graph formalism to successfully find deontic inconsistencies within the semantic information contained in sentences, and possibly in the future amongst much larger linguistic structures. The experimental system developed at the University of Westminster, currently holds most of the text of the 1984 Data Protection Act and is capable of identifying some potential conflicts of the type shown in figures 3 and 4 within the text.

Current work, which we hope will bear fruit in the near future, includes work on extending the coverage, and the dictionary of terms used by the system, the problem of anaphoric referencing, visual grammars and possibly the pre-processing of text into a form that may be accessible via the Internet, especially for revision control in multi-authored texts. It has also been proposed as an element in a much larger Management Information System [29]; and current investigation involves possibilities within the field of visual information.

References

1. Aho, A.V. and Ullman, J.D. (1972) *The theory of Parsing, Translation and Compiling, Vol: Parsing*, Prentice-Hall, Englewood Cliffs, N.J.
2. Berg-Cross, G. and Price, M. E. (1989) Acquiring and managing knowledge using a conceptual structures approach, *IEEE Transactions on Systems, Man and Cybernetics*, vol. 19:3, p.513-27.
3. Calder, J., Klein, E., and Zeevat, H. (1988) Unification Categorial Grammar : A Concise Extendable Grammar for Natural Language Processing, *COLING-88*, 83-86.
4. Clocksin, W. and Mellish, C. (1981) *Programming in PROLOG*, Springer-Verlag: Berlin.
5. Dowty, D, R., Wall, R. E., and Peters, S., *Introduction to Montague Semantics*, Reidel, Dortrecht, 1981
6. Early, J. (1970) An Efficient Context Free Parsing Algorithm, *Communications of the A.C.M.*, v.14,453-460.
7. Ellis, G. (1993) Efficient Retrieval from Hierarchies of Objects using Lattice Operations, in G.Mineau, B.Moulin, J.F.Sowa, eds, *Conceptual Graphs for Knowledge Representation: First International Conference on Conceptual*

Structures, ICCS'93, Quebec City, Canada, August 4-7, 1993 Proceedings, Springer-Verlag: Berlin Heidelberg.

8. Fargues, J., Landau, M-C., Degaud, A. and Catach, L. (1986) Conceptual Graphs for semantics and knowldge processing, *IBM Journal of Research and Development*, v.30:1, January 1986, p.70-79.

9. Fillmore, C. J., "The Case for Case" In *Universals in Linguistic Theory*. Edited by E. Bach and R. T. Harms. Holt, Reinhart and Willson, New York, 1968.

10. Knuth, D.E. (1965) On the translation of languages from left to right, *Information and Control*,v.8,607-639.

11. Konstantinou, V., Sykes, J. and Yannopolous, G. (1992) Legal Reasoning Methodology: The missing link, *Law, Computers and Artificial Intelligence*, Vol.3:1, p29-46.

12. Landau, M-C. (1990) Semantic representation of texts, *COLING-90 Proceedings*, Vol 2, p239-244.

13. Levinson, R. and Ellis, G. (1992) Multilevel hierarchical retrieval, *Knowledge-Based Systems*, v.5:3,233-244.

14. Minski, M., "A Framework for Representing Knowledge." In *Mind Design*. Edited by J. Haugeland. Cambridge, MA: The MIT Press, 1981, 95-128.

15. Nogier, J-F. and Zock, M. (1992) Lexical Choice as Pattern Matching, *Knowledge-Based Systems*, v.5:3,200-212.

16. Pau, L. F. and Nielsen, S.S. (1989) Conceptual Graphs as a visual language for knowledge acquisition in architectural expert systems, *SIGART Newsletter, Special Isssue on Knowledge Acquisition*, no. 108, April 1989, p.151.

17. Pareira, F.C.N. and Warren, D.H.D. (1980) Definite Clause Grammars for language analysis - a survey of the formalism and a comparison with augmented transition networks, *Artificial Intelligence*, v.13,231-278.

18. Pollard, C. and Sag, I.A. (1987) *Information Based Syntax and Semantics*, CLSI Lecture Notes 13, Stanford: CLSI

19. Quillian, M. R. (1968) "Semantic Memory", In Semantic Information Processing, Edited by M. Minsky. Cambridge, MA: The MIT Press, 227-270.

20. Shank, R. C., *Conceptual Information Processing*. North Holland, Amsterdam, 1975.

21. Shank, R. C., and Ableson, R. *Scripts, Plans, Goals and Understanding*. Hillsdale, NJ: Lawrence Erlbaum Associates, 1997.

22. Sheiber, S.M. (1988) *An Introduction to Unification Based Approaches to Grammar*, CSLI Licture Notes 4, Stanford:CLSI.

23. Slagle, J. R., Gardiner, D. A. and Han, K. (1990) Knowledge specification of an expert system, *IEEE Expert*: hp2.p:4:ehp2., 20-38

24. Sowa, J. F.(1981) A conceptual schema for knowledge based systems, *Proceedings of the Workshop on data Abstractino, Databases, and Conceptual Modelling*, ACM, p193-195.

25. Sowa, J. F. 1984. Conceptual Structures: Information Processing in Mind and Machine, Addison-Wesley, Reading, MA.

26. Sowa, J.F. and Way, E.C. (1986) Implementing a Semantic Interpreter Using Conceptual Graphs, *IBM Journal of Research and Development*, v.30:15,57-69.

27. Sowa, J. F. 1994 Knowledge Representation: Logical,Philosophical, and Computational Foundations, Preliminary Edition, ICCS'94.

28. Sowa, J. F. (2000) *Knowledge Representation: Logical, Philosophical and Computational foundations*, Brooks/Cole.

29. Sykes, J. T., Konstantinou, V. and Morse, P. L. R, "An incremental parser for Conceptual Graphs", Proceedings of the 3rd International Conference on Conceptual Structures, University of California, Santa Cruz, August 1995.

30. Velardi, P., Pazienza, M.T., and DeGiovanetti,M. (1988) Conceptual Graphs for the Analysis and Generation of Sentences, *IBM Journal of Research and Development*, v.32:2, 251-267.

31. Way, E.C. (1992) Metaphor as a mechanism for reorganizing the type hierarchy, *Knowledge-Based Systems*, v.5:3,223-232.

32. Zeevat, H., Klein, E., and Calder, J. (1987) An Introduction to Unification Categorial Grammar. in *Edinburgh Working Papers in Cognitive Science, Vol. 1: Categorial Grammar, Unification Grammar and Parsing*, ed. by N.J. Haddock, E. Klein and G.Morril.

33. Zeevat, H. (1998) Combining Categorial Grammar and Unification in *Natural Language Parsing and Linguistic Theories*, ed by Reyle, U. and Roher, C., D Reidel, 202-229.

Case-based Induction-tree Analysis in Diagnostic Protocol Development

Authors:
Dr Andrew Harvey, Dr Joseph Devlin
(Pinderfields and Pontefract Hospitals NHS Trust)
Dr Roy Johnston
(IMS Maxims plc, Dublin)

Abstract

An induction-tree procedure, based on a tool developed in the context of the INRECA case-based reasoning EU project(1), has been used as an aid to adjusting weighting factors for use with medical diagnostic indicators in the Rheumatology domain. On a pilot sample the hit-rate of successful diagnosis using an IT based protocol diagnostic procedure, as compared to expert diagnosis by experienced consultants, was increased substantially. This indicates that the accuracy of weightings to be attributed to various indicators in various situations may be significantly improved by the use of inductive analysis of a case-base, leading to increase in appropriate referrals by GPs and accuracy of diagnosis by junior medical staff, with reduction of workload on consultants, whose experience becomes increasingly embedded in the routine of the diagnostic protocols, and a general increase in the quality of care delivered.

1 Introduction

What follows is not a report of a 'planned scientific experiment'; it is rather a record of how a diagnostic protocol was pragmatically improved using the induction engine as a tool for abstracting and structuring expert diagnostic experience. We would be interested in exploring the utility of this process in other diagnostic domains, and perhaps coming up with procedural rules for the use of inductive evidence.

1.1 Background

We have been developing in the Pinderfields and Pontefract Hospitals NHS Trust a diagnostic protocol in support of the procedure for the referral by GPs of patients to specialist consultants. The selected pilot domain was that of Rheumatology, which included initially the following conditions:

A. Rheumatoid Arthritis (RA)
B. Osteo Arthritis (OA)
C. Gout
D. Fibromyalgia (FM)
E. Polymyalgia Rheumatica (PMR)
F. Other Inflammatory Arthritis (IA)
G. Ankylosing Spondylitis (AS)
H. Cervical/lumbar spondylosis (C&L)
I. Local or regional non-articular pain syndrome (L&R)

The foregoing conditions were identified by the June-September 1999 version of the diagnostic protocol at a somewhat unsatisfactory 'hit rate', as measured against subsequent expert consultant diagnosis. Of the 45 cases, only 13 were 'hits', with 6 'maybes' and 26 wrong.

We summarise this as follows:

Condition	FM	IA	LR	OA	Gout	CL	RA	WRULD	AS	Arthalgia
Actual working diagnosis	8	8	7	4	3	8	5	1	?	1
Protocol suggestion	1	9	4	-	2	1	11	-	17	-
Correct hits by protocol	1	4	1	-	1	1	4	-	?	-

The June 1999 version of the protocol had the following parameters and values, the values being given weightings as regards their significance in the contexts of various conditions, on the basis of expert opinion.

Symptoms and Signs: 10 values
Age at onset: 4 values
Presentation: 3 values
Time Course: 4 values
Pattern of Joint Involvement: 8 values
Other features: 8 values (this is a list of additional possible signs and symptoms)
Investigations: 8 values (this is basically a classification of the results of lab tests)
Treatment effects: 4 values (responses to various trial treatments)
Other history: 4 values reflecting family histories of named conditions
Past history: 2 values reflection patient history of specific conditions

Each value of each parameter was allocated a weighting which was condition-specific, estimated on the basis of expert experience. For example the symptom 'single swollen joint' had a high weighting in favour of conditions B and C, and a high contra-weighting for conditions D, E and H.

IMS have been gaining some diagnostic experience in non-medical domains using an induction engine which was developed in the context of the EU-funded INRECA project(1), the objective of which was to develop some tools originating in the 'artificial intelligence' research domain for use in a commercial environment. We felt that the utility of these tools might be worth exploring in a medical domain.

2 Inductive Analysis of Case-base

In IMS we took the 45 active cases which the Trust group had on record, and ran them through the induction engine. This built a decision tree on the basis of analysis of the values of all the factors, which determined the values of the target outcome, i.e. the expert diagnosis as given in the final column of the case-record array. This took no account of the expert-allocated weightings of the values; equal weights were simply allocated to all factors, the induction engine was run to generate the tree, in the expectation that weightings for the values would suggest themselves in the light of experience. The result was as follows:

According to the induction-tree analysis, the first factor determining the outcome is 'pattern'. The largest group contained 22 cases having the value 'small + large joints'. In this group, the next most important determinant was 'time-course'.

There were 17 cases for whom 'time-course = 'continuous'. In this group, if there is no 'pain', there is in all cases a positive 'esr', of a wide range of values, leading to an RA diagnosis, except in one case, where the 'esr' is high at 84, and there is no 'stiffness'; here there is a possibility it could be RA or IA. (In the later group in fact we combine these two conditions, given that they are closely related).

There are 4 cases for whom 'time-course' is 'intermittent'; for these if the onset is sudden or rapid we have IA, and RA if gradual.

The one remaining case has timecourse = s12-14 and this is diagnosed as gout.

The next largest 'pattern' group is '2-4lj' containing 9 cases. 'Stiffness' comes next. Of these 7 have no stiffness; these are diagnosed as AS if 'pain' is 'none' or 'neckback', and as IA if 'pain' is 'general'. If 'stiffness' is 'local' we get LR, and if 'stiffness' is 'general' we get IA.

There are also 9 cases in the 'pattern' = 'diffuse' group. This group next splits by age. If age <40 there are 5 cases diagnosed as AS.

There are 2 cases in the group 40-60; one with 'pain' = 'neckback' is AS, the other with 'pain' = 'general' is RA.

There are 2 in the group >60; .joint' = 'none' gives FM and 'joint' = 'multiple' gives PMR.

Next we have 6 cases in the 'pattern' = '> 5 small joints' group. For these we look next at the 'timecourse'; one 'intermittent' is gout, two 'sl<48' are RA; there are 3 for whom 'timecourse' = continuous; of these 2 are RA if there is no 'stiffness', and maybe IA or RA if there is local stiffness.

Then finally there are 2 cases for whom the 'pattern' = '1 large joint'; for these if joint = none we have LR, and if joint = single we have OA.

This was a crude first pass, and it showed up quite clearly that the first factor to go for was the 'pattern', and that diagnoses were usually feasible using a relatively small set of indicators, with the other factors remaining in the background, for use in confirmation, for reassurance of the consultant that the initial diagnosis was correct.

3 Revision of Protocol

In the light of the foregoing, the GP referral protocol was substantially revised, with much more attention being given to the 'pattern' as a discriminant, and the original over-complex 'signs and symptoms' parameter subdivided into groups of related sets of parametric values. The revised protocol contains the following parameters:

Pain (4 values)

Swelling (4 values)

Stiffness (5 values)

Pins and Needles, Tingling, Numbness (PNTN) (3 values)

(The foregoing are a subdivision of the earlier parameter 'signs and symptoms')

Joint Pattern (9 distinct values, increased from 8)

Tender points (3 values)

Muscle (2 values)

Each value of each parameter is associated with a set of weighting factors reflecting its significance in the context of each of the specified diagnostic outcomes.

Some re-grouping and re-naming of the diagnosed conditions was also done to give a broader grouping of diagnoses, this was carried out without any expectation that a scientific 'before and after' study might be undertaken.

There is however the correspondence is as follows:

RA and IA are grouped as IA (inflammatory arthritis)
Gout becomes CA (Crystal Arthritis)
AS (ankylosing Spondylitis) remains the same.
PMR becomes CTD (connective tissue disease)
OA is re-named OST (osteoarthritis)
C&L becomes C/LS (cervical/lumbar spoldylosis)
FM (fibromyalgia) and L&R (local or regional non-articular pain syndrome) are grouped as NAR (non-articular rheumatism)
and a new condition N/R is identified (neuropathy/radiculopathy).

The foregoing revised diagnostic protocol was constructed by the medical experts in the Pinderfields and Pontefract Hospitals NHS Trust, using the insights derived from the earlier inductive analysis carried out by IMS.

4 Performance of Revised Protocol

The 'hit-rate', using the revised protocol on the sample of cases available to date, after adjusting the weightings and parametric structure on the basis of the inductive analysis, can in principle be arranged to be 100% by suitably adjusting the weightings, using an iterative procedure with the inductive analysis. This however should however in no sense be construed as 'planned experiment', with a 'before and after' procedure on a comparable basis.

In order to measure the effectiveness of the first iteration of the inductive procedure, we took the cases which had been 'misses' under the old protocol, and ran them through the new one. The results for the first five were as follows:

Case old protocol expert diagnosis (old schema) new protocol expert diagnosis (new schema)

Case	old protocol	expert diagnosis (old schema)	new protocol	expert diagnosis (new schema)
1	AS	FM / (C&L)	NAR	NAR
2	Gout	OA	OST	OST
3	AS	L&R	NAR	NAR
4	AS	L&R	NAR	NAR
5	RA	FW	NAR /(AS)	NAR

For this preliminary small batch the new protocol now works perfectly.

A further group of 16 of old-protocol 'misses' was re-run under the new protocol. We were unable to correlate the re-runs case by case as above, but the overall picture is as follows:

Condition:	NAR	CLS	OST	RA	PMR	AS	IA	CA
diagnosis	6	4	1	.5	.5	-	3.5	.5
protocol correct	6	1.5	.5	-	-	-	3	.5
protocol incorrect	-	-	1.5	-	-	2.5	.5	-

The occurrence of 'half-values' is due to the existence of doubt as between two competing diagnoses. If two protocol-suggested rankings are the same or one unit apart we take them as being both equally plausible alternatives.

This result suggests that in the next iteration of the induction engine, with the new protocol, we will look closely at the branching of the tree leading to AS, OST and IA, and consider giving higher weightings to the secondary and tertiary diagnostic factors which pick these out. Unfortunately we have been unable to do this within the available timeframe.

The main difference between the old and the new protocols was the attention given by the Rheumatology experts to the details of the values of the parameters, based on the assessment of their importance, as derived from the first run of the induction engine.

There were also indications from the inductive analysis which suggested how the weighting factors should be revised. This processing of the output of the inductive analysis was subject to selective assessment using consultant experience. In other words, the role of the induction engine was as a tool to provide previously unavailable suggestions in support of the process of expert adjustment of the structure of a diagnostic protocol, and of the weighting factors assumed within it.

The induction tree is in no sense a 'substitute' for expert diagnostic procedures; rather it is a support system for the structured quantitative abstraction of expert experience, which in the case of Rheumatology in Pinderfields and Pontefract Hospitals NHS Trust has proven capability to deliver improved diagnostic protocols and embed it within the technology for ease of dissemination

It is to be expected that as new cases arise, even though the weightings may have been adapted to give a near perfect result for the cases to date, as the size of the sample population increases more 'misses' will again begin to appear. When these accumulate to some defined level, there will be a re-run of the induction engine, generating a further revision initially of the weightings, then if necessary of the protocol structure. By repeating the process successively, we in effect formalise an 'organisational learning by experience' procedure.

5 Conclusions

The inductive analysis of a case-base, with generation of a decision tree derived from a set of values of diagnostic parameters, to attain an outcome consisting of an expert diagnosis achieved independently by an experienced consultant, can provide a useful indicator of the relative values of diagnostic indicators, and can provide a tool for the development of improved diagnostic protocols, for use in training and referral.

Further work is necessary, primarily in the elucidation of explicit procedures for translating induction-tree information into diagnostic protocol structures and weightings. There is also a need to extend this procedure into other diagnostic domains, and the elaboration of rules for recognising how such domains should be delimited, in such a way as ensure that the inductive tree serves as a clear indicator. Any attempt to make the domain too large would result in a decision tree of such complexity that the elucidation of valid weighting rules would be difficult.

There is also scope for elucidating weighting factors for allocation to parameters, as distinct from weighting factors for condition-identification to values within each parameter.

The definition of what constitutes a 'parameter' and what constitutes a 'value within a parameter' is another grey area, currently a matter for negotiation between the 'induction tree specialist' and the 'medical diagnostic specialist'. We do not yet pretend to have answers to any of these questions. We have simply reported that we have found the inductive analysis useful in improving diagnostic referral protocols in one specific delimited domain, and that as a consequence consultants' workload could be reduced, and the accuracy of GP diagnosis improved.

Reference

1. Bergmann R. et al. Lecture Notes in Artificial Intelligence, 1612, Developing Industrial Case-based Reasoning Applications - the INRECA methodology. Springer Berlin, 1999; see in particular p159 for examples of applications of the use of inductive analysis in engineering system fault diagnosis.

Acknowledgements:

The work described here was supported by the CEC ESPRIT Programme.
through project 27066 MEASURES (Metrics to Support Integration and Evolution of Healthcare Systems) which completed in July 2000.

The authors wish to thank Angela Millett of Pinderfields and Pontefract Hospitals NHS Trust who has been extremely helpful with the abstraction of case-base data into a form which made this paper possible at relatively short notice.

Smart Radio - Building Music Radio On the Fly

Conor Hayes & Pádraig Cunningham
Computer Science Department
Trinity College Dublin
conor.hayes@cs.tcd.ie
www.cs.tcd.ie

Abstract. This paper describes the development of a networked music application at Trinity College Dublin. Smart Radio is a web based client-server application which uses streaming audio technology and collaborative recommendation techniques to allow users build, manage and share music programmes. While it is generally acknowledged that music distribution over the web will dramatically change how the music industry operates, there are few prototypes available to demonstrate how this could work in an managed way. The Smart Radio approach is to have people manage their music resources by putting together personalised music programmes. These programmes can then be swapped using techniques of collaborative recommendation to find similarities between users. The smart radio system currently runs within the Computer Science Intranet with permission from the Irish Music Rights Organisation (IMRO). It is a prototype system for an "always on" high bandwidth Internet connection such as ADSL.

1. Introduction

The availability of high quality digitally compressed music on the Internet has caused waves in the music distribution industry. The sale of music has been a highly centralized activity involving the record companies who record artists, distributors who ship music media items such as CDs, and retails outlet who make music artefacts available to the consumer. CD prices are generally fixed across geographic borders (government taxes and shipping costs differentiate one region from another). Up to recently the basic purchase unit of music has been a collection of tracks by a single artist or band. Though compilation albums of mixed artists are available, their number are generally low compared to single artist items. However, the music industry does not offer the facility to purchase in-store personally built compilations even though this activity is commonly carried out by consumers using blank tapes and, of late, recordable CDs. A common off-shoot of this activity is that home-made compilations are swapped or given as presents despite the fact that this contravenes the artist's copyright. With the advent of MP3 compression techniques and the ease with which people can now send data over the Internet, near CD quality digital tracks are suddenly easily available for millions of people to download and store locally. The piracy aspect of this activity is greater than 'home taping' in that an unregulated distribution channel (The Internet) had been put in place. The music industry has little agreement on issues of digital authentication, or on how to decentralise their sales operations to take into account the new paradigm of personalised delivery over a network. Furthermore, there is continuing debate on the payment system suited to music download.

The Smart Radio Project was initiated to address these issues. It does so by assimilating two paradigms:

- The regulated distribution of music on the web
- The service of personalised radio programming

In the section 2 of this paper we will describe our motivation for using streaming music technology as opposed to downloadable MP3s. Section 3 will address how we use collaborative recommendation techniques to provide in effect a personalised radio service to users. We argue that the most useful unit of recommendation is a music programme, a user built compilation. We show how user feedback occurs at two levels of granularity - at the track level and at programme level. Further to this we discuss how the Smart Radio recommendation system encourages community participation by allowing users know who their most consistent neighbours are. In this way, a recommendation system encourages rather than replaces social processes [1].

Sections 4 and 5 describe the system architecture and operation. Sections 6 and 7 discuss proposed payments systems, and the potential for a Smart Radio system in conjunction with higher bandwidth technologies such as Asymmetric Digital Subscriber Lines (ADSL).

2. Music Formats

Smart Radio uses streaming audio technology developed by Real Networks[1]. The architecture is not tied to this technology but at the time of development Real Networks provided the most popular streaming client player.

Streaming technology was chosen (as opposed to a complete download model) for a number of reasons. The service paradigm that we chose to examine was that of personalised radio. Since music radio is well understood concept, by tying together the concept of a music programme with a personalised recommendation system, we could provide an intuitive stream of music service. For example, if the basic unit of recommendation is an mp3 file, the time involved in downloading several of these to a hard drive and then setting up a play-list is quite considerable. With the streaming technology used by Smart Radio, a user can login, choose a recommended programme or a favourite programme, and have a music programme stream to the desktop almost immediately. As well, as the benefits of immediate delivery, we considered streaming technology to be a useful solution to the piracy

---

[1] http://www.realnetworks.com

issue. The streaming audio files are not saved to the user's hard drive, or posted on the internet for illicit distribution. Another benefit of a service where there are no downloadable files is that the user has a roaming profile and his music programmes and recommendations are available wherever he logs on. Though disk space has become cheaper, there is considerable management involved in holding a large digital library on a local machine. If each file has been paid for then back-ups need to be made in case of disk failure.

3. Collaborative Recommendation

Collaborative Recommendation techniques have proven a attractive methodologies for making recommendations where there are few or unhelpful features by which to categorize assets. Some of the earliest work in this area was done in the area of music recommendation [2]. The techniques have come into their own with the growth of e-commerce stores such as Amazon.com. The basic collaborative recommendation model supposes that a person's rating of a new resource can be predicted by looking at how other similar users have rated the same resource.

The system makes a correlation between a user and other users whose profile are similar according to a similarity metric. From this correlation it then calculates items that the group have in common, which are missing from the active user profile. Smart Radio uses a measure of co-linearity as the correlation between the active user and the user population. This measure, known as the Pearson R coefficient, has been reviewed and found to be a satisfactory indicator of user similarity [3].

The technique used in Smart radio is to choose the top *n* neighbours for each user and then use these neighbours as the source of recommendations. For items in the neighbourhood that the active user has not yet rated predictions are made using a weighted average of deviations from each neighbour's mean vote:

$$p_{a,i} = \bar{r}_a + \frac{\sum_{u=1}^{n} (r_{u,i} - \bar{r}_u) * w_{a,u}}{\sum_{u=1}^{n} w_{a,u}}$$

Equation 1: Predicting user a's rating for item i

pa,i is the prediction for user a for item i. $r_{u,i}$ refers to user u's rating for item i. $\bar{r}_a$, $\bar{r}_u$ refers to the rating average calculated for user *a* and *u* respectively. *n* is the number of neighbours and $w_{a,u}$ is the weight between user *a* and neighbour <u>u</u> .

This weight is given by the pearson correlation coefficient between user a and user u:

$$w_{a,u} = \frac{\sum_{i=1}^{m} r_{a,i} * r_{u,i} - \frac{\sum_{i=1}^{m} r_{a,i} \sum_{i=1}^{m} r_{u,i}}{m}}{\sqrt{\left(\sum_{i=1}^{m} (r_{a,i}^{2} - \frac{(\sum_{i=1}^{m} r_{a,i})^{2}}{m}) \right) \left(\sum_{i=1}^{m} (r_{u,i}^{2} - \frac{(\sum_{i=1}^{n} r_{u,i})^{2}}{m}) \right)}}$$

Equation 2: Pearson correlation coefficient

In equation 2, m refers to the number of items the two users have in common. In order to ensure that correlations are not being calculated over a small number of common items a further weight is applied. With our current user population we found it was necessary to have rated 20 items in common before good recommendations were being made. Therefore, if a pair of users has less than 20 items in common the correlation obtained by the Pearson measure is devalued by m/20.

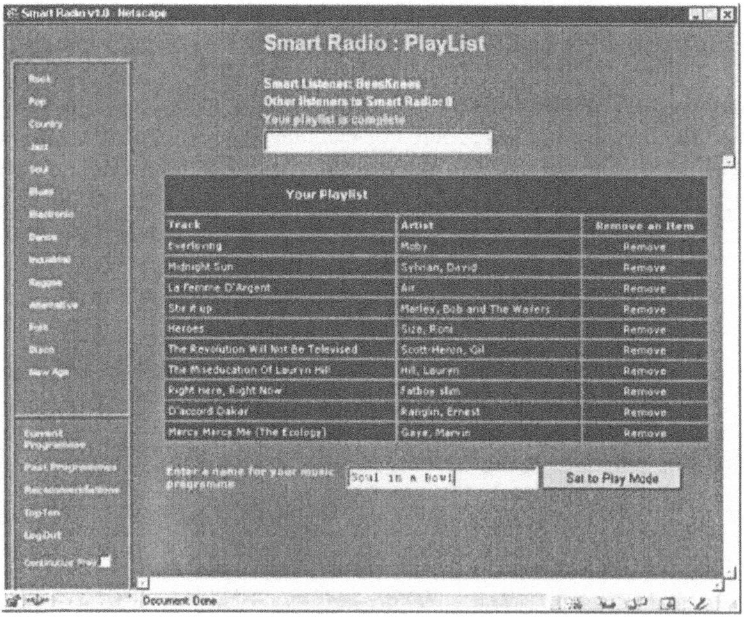

Figure 1: Naming a recently built play list

User ratings are gathered using explicit and implicit user feedback. Users can explicitly choose to rate individual track items or individual programmes on a scale of 1 – 5, where 5 is the top score. Implicit feedback is important for the less interactive or casual user [4]. The reasoning behind this is that the casual user wants to interact as little as possible with the system and will intervene only when the system gets things badly wrong. In terms of implicit feedback the system allocates a score of 4 to recommended programmes and constituent items that have been saved to a user's profile, or to programmes that have been built from scratch by the user. Programmes can be put together in a matter of seconds by simple clicking on the desired music item and adding it to the current play list (See figure 1). Music items are indexed currently by genre, which is not satisfactory since the genre feature is not finely grained enough to capture the nuances within music types. Further research is planned on how to allow user indexing of music assets. Once a programme has been built it can be played immediately and is automatically saved to the users profile for future retrieval. Programmes that are played more than three times are awarded the top score of 5, even though the average rating of constituent items may be lower. Our theory is that a well chosen collection of music has greater value than the sum of its constituent items. For one thing, there is some work involved in putting together a programme so there is some value in choosing something "off the shelf". For another, a collection of music may contain the difficult to quantify feature of "mood" which depends on the collected items being played together. This feature is apparent where users amend their ratings for individual items as they appear in different programmes. Figure 2 illustrates an excerpt for the programme *mellow and jazzy* in which the user *cchayes* has rated four out of the five shown items. If cchayes chooses *mellow and jazzy* again he will be shown his ratings for the individual items within the programme and he may recast his vote. This facility is important because music taste does shift, and user profiles will have to move to reflect this. It is entirely probable that a user will cease to become a recommender in one neighbourhood only to have moved to another.

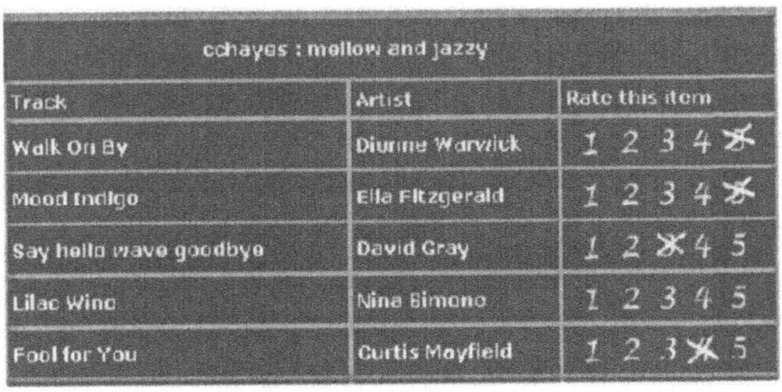

cchayes : mellow and jazzy

Track	Artist	Rate this item
Walk On By	Dionne Warwick	1 2 3 4 5
Mood Indigo	Ella Fitzgerald	1 2 3 4 5
Say hello wave goodbye	David Gray	1 2 3 4 5
Lilac Wine	Nina Simone	1 2 3 4 5
Fool for You	Curtis Mayfield	1 2 3 4 5

Figure 2: a portion of play list entitled *mellow and jazzy*

4. Smart Radio Architecture

Smart Radio is a web based client server application. Users must login through their browser, after which they may choose to build new programmes from scratch, play past favourite programmes, or choose programmes recommended by their neighbours. Server side programming was done with Java servlets, Java server pages and Java beans running on the servlet engine in Sun Microsystem's Java Web Server. Servlets provided an elegant *state* management facility for Smart Radio. Each user logs in and is provided with a virtual session on the Smart Radio server. Since http is a stateless protocol, Java servlets allow unique session Ids be set as cookies in the users browser. These ids are returned to the server with every http request. The session id can be validated and the resources associated with each valid id retrieved from the servlet engine's hash table. In the case of Smart Radio, each online user has a neighbourhood object and a recommendation object associated with his active profile.

This is a dynamic domain where users may seek new recommendations almost immediately once the current programme has played through. The first version of Smart Radio calculated new recommendations as they were requested by online users. A user-correlation matrix was held in memory and updated as each user required fresh recommendations. However, this method was not considered computationally or architecturally scalable. The current implementation separates the web server processes from the recommendation engine, allowing both to be scaled as necessary. A scheduler recalculates the neighbourhood for each *online* user on the half hour, and updates the database with the most current recommendations. The web server then queries the updated database for new recommendations. Currently the web server, the scheduler and the recommendation engine reside on a single machine. However, since the system is implemented in Java these modules can be easily moved to remote machines and would communicate using Remote Method Invocation.

The current database being used by Smart Radio is Microsoft Access which has proven adequate to the traffic generated within the Intranet. We will have to examine a scalable database solution if this work is to move to the Internet. Database access is handled by java database manager running on the server. The manager creates a pool of connections from which it allocates connections in a round robin fashion as requests are received from components of the system. If there are no connections available, the request is put on a queue, and assigned the first released connection.

A Real Audio Server runs in parallel with the Java Web Server. This server receives requests from the clients and streams audio pieces to the specified address. The Smart Radio Web pages are able to play the incoming stream by means of the Real Audio plug-in embedded in a HTML frame which remains constantly in place irrespective of what other pages the user requests from the Smart Radio server. In fact it is this plug-in that requests the file containing the addresses of constituent pieces of the programme ordered by the user.

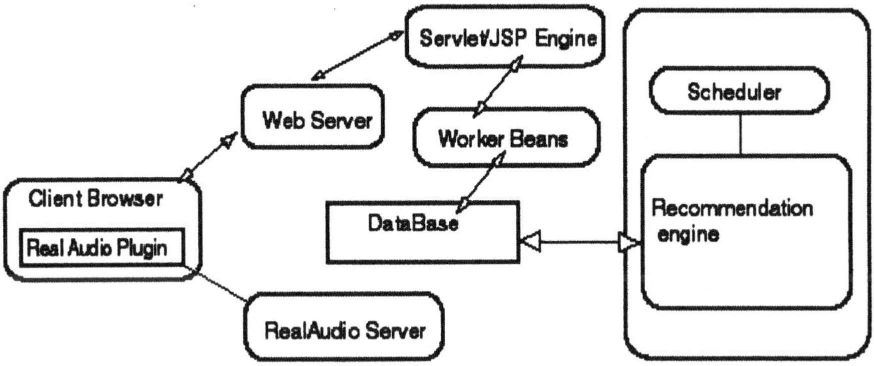

Figure 3. The architecture of the Smart Radio system

5. Receiving a programme of music

There are two stages to receiving a programme of music. The user can choose to either configure a new programme, choose a recommended programme or a past favourite. When the user indicates that the programme is ready to play (he may want to edit it before setting it to play) the browser passes the session id to the plug-in, which requests the programme associated with this session id. The server has already configured a long XML string containing the URLs of the new programmes and stored it a hash table with the session Id as key. The XML file is written according to the format of the synchronized multimedia integration language, SMIL. The server receives the request and returns the SMIL string to the plug-in which begins buffering the first music item. It is important to note that programme files are created dynamically, and do not reside on the server. The database contains the URL for each music item and SMIL servlet builds the programme on the fly. Once the SMIL file is received, a hidden java applet monitors the activity of the plug-in and using browser scripting allows the user to see the status of the music stream (i.e. whether it is buffering, which track is playing and any connection errors).

6. Payment systems and future Viability

In Ireland, terrestrial radio is license-free for the user, though some of the television license funds may supplement radio programming. In order to comply with copyright law, radio stations have to pay a royalty fee determined by the Irish Music Rights Organisation (IMRO).

The Smart Radio system may well be suited to a brokerage or middle-ware role between content vendors and listeners.

The first payment system to consider is a subscription or license model, similar to the subscriptions charged by cable companies. The subscription charge is for a service – access to streaming music as opposed to a series of audio downloads. Subscribers pay a flat fee per quarter. The brokerage firm then pays the content providers for their services. This would either be a flat fee or in proportion to the content they provide. Revenue would accrue to the Smart Broker from a proportion of the subscription fee and the direct marketing service being offered to music vendors, such as CD Now, for instance.

The second scenario involves charging per unit content received by the client. The most useful method here would be a micro-payment[2].

The client's relationship is with the Smart Radio broker. He may pay in advance into a personalised account. Micro-charges are extracted from his account per programme. An SSL connection will have to be set up to credit his account initially, but also *every time* a play-list is sent to his desk top since the client will have to be verified before charges are deducted from his account. This is difficult to scale since setting up an SSL connection has a computational over head. Certainly, a play-list is the smallest unit for which a user should be charged in a real-time environment

Of the two schemes, a periodic subscription charge may be the most suitable format since users may be hostile to the idea of paying for a single listen to an item rather than having it to download.

7. Internet Viability

Streaming Audio technology operates using TCP to establish a connection and the unreliable datagram service UDP to stream packets to the client. The unreliability of the UDP format means that packets may not arrive in time for playback due to network congestion. The problem is exacerbated by the transmission rates required to stream near CD-quality audio (even while highly compressed), approximately 128kbs. The Smart Radio system running on the computer science departmental intranet uses audio files encoded at bit rates between 44kbs and 128 kbs and occasionally encounters problems when traffic is high. On PC speakers the drop in quality caused by using a 44Kbs codec is not particularly noticeable.

---

[2] http://www.ipin.com

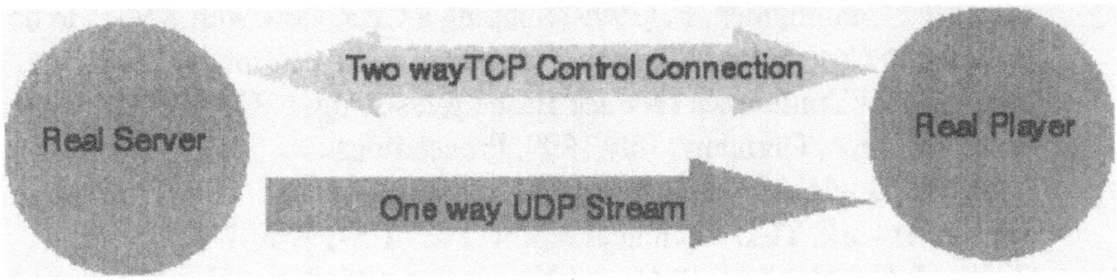

Figure 3. Streaming audio connections

However, even at this codec the transmission rate required would require most of the bandwidth available on good local telephone loop using a 56kbs analog modem. The system could run as an AM quality service on the local loop, but really would not be a viable service until internet dialup charges are free. It is for this reason we propose Smart Radio as the type of service suited to Asymmetric Digital Subscriber Lines (ADSL), a technology which is likely to change domestic use of the Internet dramatically. ADSL provides high bandwidth access to the Internet (up to 8Mbs) over existing local loop wiring. Since it uses frequency bands outside of that used for telephony (<4khz), ADSL does not have a dialup mechanism. It provides, in effect, an "always on ", dedicated line to the Internet for anyone with a telephone connection. It should be clear that these sort of conditions would prove ideal for an application such as Smart Radio.

References

[1] Hill, W., L., Rosenstein, M., and Furmas, G. (1995) Recommending and Evaluating Choices in a Virtual Community of Use. In proceedings of the conference on human Factors in Computing Systems (CHI95), 194-201, Denver, CO, ACM Press

[2] Shardanand, U., Maes, P. (1995) Social Information Filtering: Algorithms for Automating "Word of Mouth", In Proceedings of the Conference on Human Factors in Computing Systems (CHI95), 210-217, Denver, CO, ACM Press

[3] Billsus, D., Pazzani, M.J. (1998) Learning Collaborative Information Filters, in *Proceedings of the International Conference on Machine Learning. (ICML 98)* Morgan Kaufmann Publishers. Madison, Wisc.

[4] Loeb, Shoshana (1992) Architecting Personalized Delivery of Multimedia Information. Communications of the ACM/December 1992/Vol.35. No12

[5] Balabanovic, Marko; Shoham Yoav; (1997) Fab: Content-Based, Collaborative Recommendation. Communications of the ACM, March 1997, Volume 40, Number 3

[6] Konstan J.A., Miller B.N, Maltz D., Herlocker J.L., Gordon L.R., Reidl J.; (1997) Grouplens: Applying Collaborative Filtering to Usenet News. Communications of the ACM, March 1997-Volume 40, Number 3

[7] Hayes, C. Cunningham, P. (1999) Shaping a CBR view with XML - to be included in Case-Based Reasoning Research and Applications. Third International Conference on Case-Based Reasoning, ICCBR-99, Monastery Seeon, Munich, Germany, July 1999, Proceedings

[8] Hayes, C. Doyle, M. Cunningham, M. (1998) CBR Net :- Smart Technology over a Network. TCD Technical report TCD-CS-1998-07

[9] Hickie, J. Cunningham, P. Neural Networks for Collaborative Filtering: A Case Study in Web Advertising.

[10] Smyth, B., Cotter, P., and O'Hare, G. M. P. (1998) Let's Get Personal: Personalised TV Listings on the Web. 9th Irish Conference on Artificial Intelligence and Cognitive Science. Dublin, Ireland.

[11] Wilke, W. Lenz, M. Wess, S. (1998) Intelligent Sales Support with CBR in Case-Based Reasoning Technology. Lenz et al. Springer-Verlag Lecture Notes in Artificial Intelligence 1400 (1998)

SESSION 3

ENGINEERING APPLICATIONS

TOWARDS INTEGRATED ONLINE SUPPORT FOR FIELD SERVICE ENGINEERS IN A FLEXIBLE MANUFACTURING CONTEXT

F Coenen*, P Leng*, R Weaver and W Zhang***

*Department of Computer Science
The University of Liverpool,
Liverpool L69 7ZF, UK
Phone: +44 (0)151 794 3798
Fax: +44 (0)151 794 3715
Email: {f.coenen, phl, wzhang}@csc.liv.ac.uk

**Stoves PLC
Stoney Lane, Prescot
Merseyside L35 2XW, UK

Abstract

Providing effective and efficient support to maintenance engineers is a significant issue facing industries operating in a flexible environment. In this paper we present a system framework, which integrates various techniques of Expert Systems, such as Model-Based (MB), Rule-Based (RB), and Case-Based (CB) reasoning methods, to supply on-line maintenance and diagnostic support. The proposed framework is generic and its usage is illustrated using a "white-goods" manufacturing example. The paper includes a review of the application domain, the identification of the particular research issues to be addressed, a description of the system architecture and its components, and a discussion of related issues.

1 Introduction

Conventional manufacturing processes deliver a range of products from a catalogue which changes only infrequently. In fast-moving industries, this leaves manufacturers constantly vulnerable to competition which identifies new market trends and supersedes their current product range. One response to this is the flexible manufacturing paradigm, in which manufacturers establish design and production processes that enable them to deliver new models to meet individual customer requirements as they arise. The greater the variation of product within this process, the greater is the ability of the manufacturer to satisfy customer needs and resist competition. An illustration of flexible manufacturing is the production of domestic cookers by Stoves PLC. The Stoves manufacturing process can deliver over 6000 versions of its cookers, and does so to produce customised models in production runs which are often in single figures.

This kind of highly flexible, demand-driven process clearly imposes quite different requirements of information management than does conventional manufacturing. A particular problem arises in meeting the requirement to provide after-sales service. Field engineers may expect to be faced constantly with the need to provide service for models which they have never previously encountered, and for which there is no case history. To meet this need, they require customised product information and, perhaps, training.

Producing conventional service manuals and other product documentation for each model variant could, clearly, impose unacceptable strains on the production cycle, and the resulting volume of documentation is likely to be quite unmanageable for engineers in the field. A partial solution, used by Stoves, is to prepare support documentation in the form of a CD which is updated periodically and issued to its service agents.

The essence of flexible manufacturing is the meeting of current demand. Somewhat similar problems are encountered, paradoxically, by the service needs of products from the opposite end of the manufacturing spectrum, which are extremely stable and have a very long lifetime. Here the problem arises because the product, over its lifetime, may receive continual modifications to implement new customer needs and incorporate updated components. The result again may be a product model which is essentially unique and, moreover, has been created by a very attenuated and possibly poorly-documented ad-hoc design process.

What is required, in both cases, is a process for the creation of product information which exactly mirrors the process by which the product itself was created. To meet the demands for field service, this information will need to be produced within the existing production timescale, and be available immediately as required by service personnel.

In this paper we describe a research project the aim of which is to address this issue. The goal of the work is a system which will generate and deliver to field service engineers, via the Internet, the information they need to provide service, and diagnose and correct faults for products, created by the processes we have outlined above. The project is supported by the DTI under the Foresight 'Link' programme, and is being carried out in collaboration with Stoves PLC and other industrial partners.

The rest of the paper is arranged as follows: 2. System Scenario, which identifies the basic entities and information involved in the proposed system; 3. Related Work, which reviews the previous research and development in related fields; 4. System Architecture, which specifies the system processes, system components and their relationships, and underlying data structure; 5. Discussion, which considers some of the key research issues identified in the project; and 6. Conclusion, which summaries the main findings and future work.

2 System Scenario

In the delivery of field service information, three key entities are involved (see Figure 1). First, the service agent in the field will hope to obtain information on demand, with minimal delay, after quoting only a brief product reference, ideally a unique identification found on the product. This information will be delivered by the product manufacturer and/or the parts suppliers. In a conventional environment, it would be possible for the relevant data to be extracted from the product catalogue directly. In the flexible context we are considering, however, it will be necessary to create the information required by reference to the product design data and, thence, to data relating to the components of the product. The kinds of information delivered could be very diverse; we will briefly consider some of the more important categories.

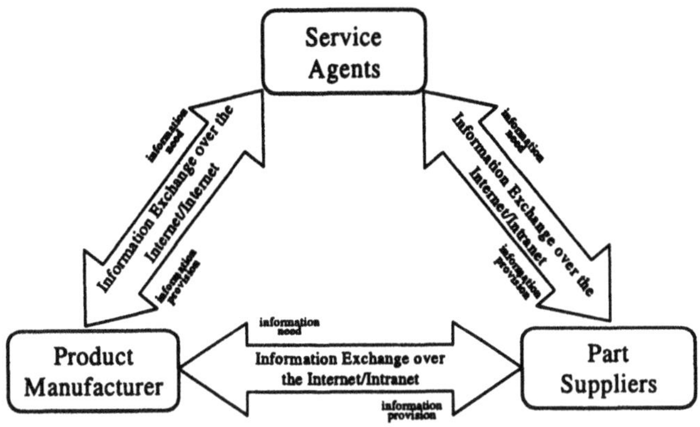

Fig 1: System Scenario

2.1 Textual and other descriptive data

Usually, simple text descriptions will be a central feature of the information provided. In an online context, the presentation of text can be augmented and structured using hypermedia techniques [5], and the information incorporated may then include other non-textual data, including photographs, video clips, and animations. In an engineering context, detailed product information in the form of plans, engineering drawings, and design data will also be essential, possibly including access to a product database.

2.2 Product Modelling

Even when dynamically generated and augmented by non-textual information, product documentation is essentially an organisation of data into a convenient and manageable form. Product modelling goes beyond this, in presenting information on the appearance and behaviour of the product in a form which allows the user (in this case, the service agent) to interact with the description: for example, to configure the model with locally-applicable data, to conduct experiments, and to focus on specific features. Virtual Reality (VR) languages, for example VRML [4] and Java3D [14], allow product models to be built which incorporate 3D graphics and enable 'fly through' simulation to illustrate behaviour.

2.3 Expert knowledge

Much of the information used by service engineers, as with other skilled professionals, may take the form not just of factual data but also of accumulated expertise. As in other domains such as medicine in which diagnosis is central, Knowledge-Based Systems may be expected to have a role here. Production rule systems [3] have been favoured by the Expert System community for many years, particularly for diagnostic systems which may often be represented readily in terms of 'if-then-else' rules, and there continues to be active research into rule-based diagnosis in many contexts [12].

2.4 Case-based data

Another important basis for fault diagnosis in the field is the use of case data describing possible faults, symptoms, verification procedures, and repairs. Case-Based reasoning (CBR) systems [15] typically store case histories in some form of data organisation, fronted by a reasoning system. New cases are investigated by presenting relevant case data to the system which then attempts to match this with recorded histories in order to identify similar cases which may provide relevant information.

2.5 Operation

Figure 2 shows the working procedure of a system to present information of these various categories to an engineer in the field. We envisage the system being invoked by the entering of a product code by a field service engineer (Figure 2(a)). This will then produce a high level model (Figure 2(b)) of the product, comprising a hierarchical tree structure with the top-level node representing the most abstract view of the product (e.g. "cooker"). Note from figure 2(b) that some nodes are linked to a "menu". These list various "information providers", relating to the categories introduced above, that are applicable to the node. Thus, at the top level, we may have a table of part numbers, diagrammatic and textual data concerning the product as a whole, diagnostic KBS and CBR style facilities, and 3-D and virtual reality facilities. At the lower levels more detailed information providers will be supplied, such as data relating to the make-up of various sub-assemblies and further expert diagnostic information. Facilities such as on-line part ordering or stock DB inquiries may also be included.

Having produced this high level view of the product the maintenance engineer will then be able to "zoom in" on a particular branch of the tree and bring up a more detailed hierarchy (Figure 2(c)). Note that some of the information providers included here have been inherited from nodes higher up (and no longer visible) in the tree. The focussing process can then be continued until the desired level is arrived at, or a single leaf node is reached (Figure 2(d)).

3 Related Work

Each of the information categories mentioned above has been widely used in systems for diagnosis and/or explanation generation in a variety of domains. A distinction can be made between systems that are essentially descriptive, i.e. producing documentation for the target domain, and those that provide a model of aspects of the domain that can be used for experiment and/or to reason about the domain. When dealing with complex structures, however, even systems which do not attempt to simulate behaviour will require to model the target domain in some detail.

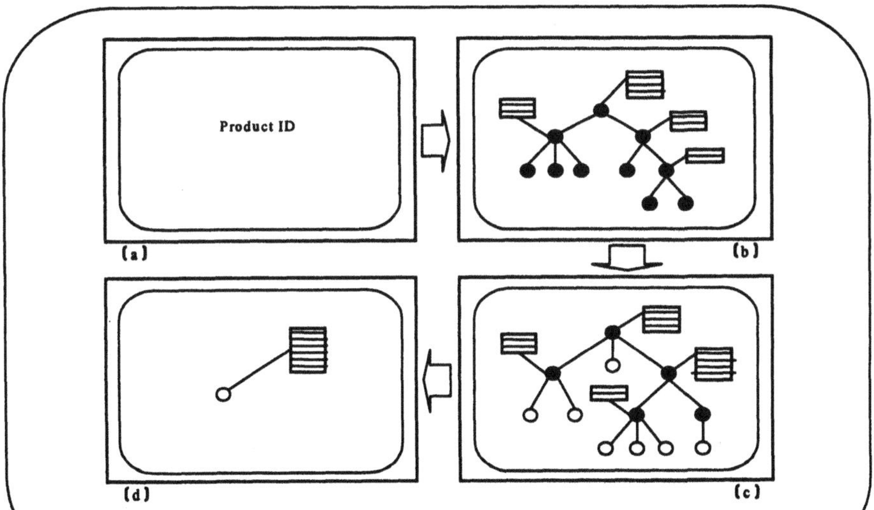

Fig 2: Working Procedure

A useful review of current work in model-based and qualitative reasoning is provided by Hunt, Lee and Price [9]. Different types of model include, in particular, structural models, which attempt to capture important physical or logical relationships of components of the target system, and functional models, which derive the operation of the system from information about the functional behaviour of its components. An example of an essentially structural model is that described by Xue, Yadav and Norrie [17] for use in building product design. This system also incorporates a knowledge-based system to generate building product descriptions automatically.

For our purposes, structural information is important: because the modelled product may be entirely new to the service agent, it will be essential to produce a model that he/she can readily relate to the real product, identifying components and their physical relationships. However, it will also be necessary, to describe operational procedures and for fault diagnosis, to build into the model representations of functional behaviour. Compositional modelling [7] derives a behavioural model from a knowledge base of the behaviour of components and their environment. Iwasaki et al [10] describe a web-based compositional modelling system, CDME, which formulates behaviour models of physical systems from domain knowledge. In CDME this knowledge is principally represented at two levels: an ontological level, which defines the vocabulary of the system, and a physical level, which describes aspects of behaviour in terms of this vocabulary.

The INT-OP project [1], concerned with the generation of operating procedures for chemical process plants, also incorporates elements of both descriptive and operational modelling. The system, CEP (Chemical Engineering Planner) produces plans which define sequences of actions to bring about specified changes in the plant. The basis of the system is a detailed model of the domain. Although the target domain of this project is more complex than the one we are considering, and the output more specialised, a number of elements of the work are relevant to our discussion. The plant model used defines a hierarchy of components and the connections between them, and a conclusion of the project was that the effort required to create this was substantial, and that tools to assist in domain development are vital. A further element of the model is a set of 'pairs', defining associations between components. The significance of this in our context is that, for example, a fault identified in one component of the system may in fact be caused by a problem arising in a different component, with no close physical connection to the one under consideration.

A number of key issues emerge from the research described in the literature. The first is the need to integrate model-based aspects of the system with other elements. Hunt et al [9] observe that even in model-based diagnosis, much of the information that is necessary is not model-based, and includes data such as failure likelihoods for different components, and heuristics for fault localisation. Secondly, the need to derive information about a target product from data relating to its components calls for the use of intelligence in this integration. Finally, both these issues point to the central role carried by the organisational structure of data in the system. This needs to incorporate not only a detailed structural model of the target domain, but also behavioural information and representations of all the different kinds of information which may be called upon by the service agent. In the following section we outline the architecture of a system to meet these requirements.

4 Framework for the proposed system

4.1 System Architecture

Figure 3 illustrates a top-level processing procedure of the proposed system.

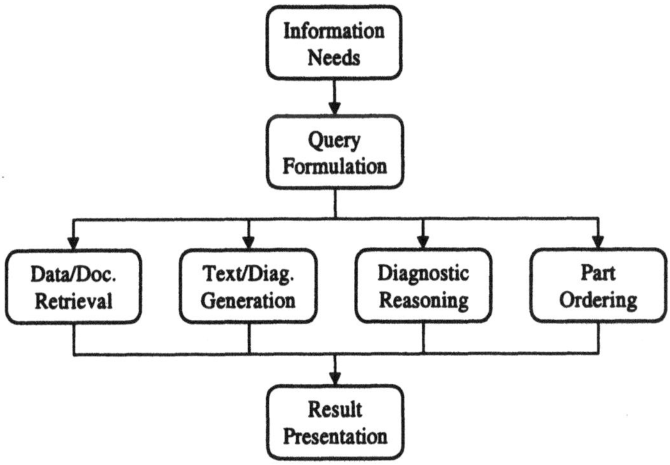

Fig 3: System Processes

Field engineers or other users raise Information Needs from their practical work. From these are formulated formal queries that the system can understand. According to the queries, one or more of the following could happen: Data/Document Retrieval, Text/Diagram Generation, Diagnostic Reasoning, or Part Ordering. Data/Document retrieval means finding some pre-stored data/documents which could satisfy the users needs; these could be of different media, such as text, image, audio, etc. Text/diagram generation uses text or diagrams for components of the product to create composite representations, for instance, 3D diagrams of products. Diagnostic reasoning addresses the other central aim of the system, to assist the engineer in fault diagnosis and repair. It may make use of techniques including case-based and rule-based systems to support this function. Finally, Part Ordering is provided to enable the engineer to obtain replacement parts when necessary. The result presentation takes the outcomes from the above processes for display to the user.

Figure 4 shows a preliminary architecture for a system to support these processes.

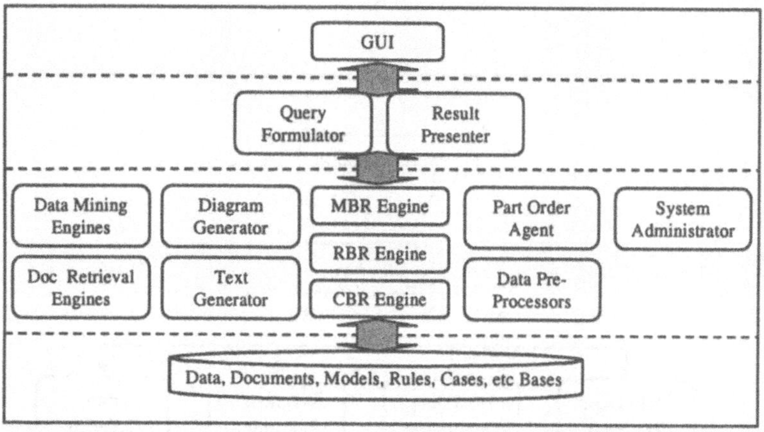

Fig 4: Integrated System Framework

The system components included in this framework can be divided into four groups: User Interface (Graphic User Interface, User Requirement Formulator, Result Presenter); Processing Agents (Data/Document Retrieval Engines, Text/Diagram Generators, Diagnostic Reasoning Engines, Part Order Agent, etc); Administrative Agents (Data Pre-processors and System administrator); and Databases (data bases, document bases, model bases, rule bases, and case bases). Most of these components have a direct correspondence to the processes of Figure 3. The implementation envisaged represents system components in the form of servers or agents operating over the Internet; this could be refined with different kinds of architecture techniques, for example, Multi-Tier Client-Server and Multi-Agent Architectures. [6, 16].

4.2 Data Model

Underpinning our architecture is a database system, which could have very complex data models, including different data types, common data, multimedia documents, knowledge bases, etc. It could also be more than one database, possibly located at different places. The underlying data structure of this information has a central role in the operation of the system. A conceptual model of this is illustrated in Figure 5, in the form of an Entity-Relationship diagram. Three principal classes of entity types are depicted: products, documents, and diagnostic models. A product unit is in the database as a hierarchy of components which have specific functional properties, and other elements (such as panel, chassis, etc) which will not contribute directly to its functional behaviour but are relevant in describing and depicting its appearance. Documentation, in the form of text and other representations, is in general associated with functional components, although some forms of documentation may be associated with the product itself. Finally, diagnostic models are included corresponding to the various different approaches identified in the architecture of Figure 4. The relationship types are identified by using the natural association amongst the above entities. The ordering function is not considered here because it can be taken to be an independent system. The product illustrated in Figure 5 is an example of 'white goods' (a cooker), but the analysis could equally be applied to other categories of product.

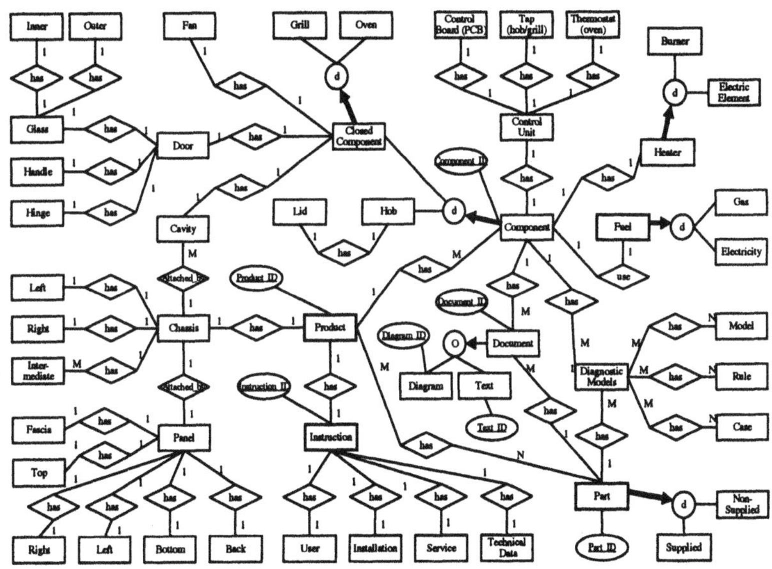

Fig.5: ER Diagram

5 Discussion

The key issue identified in the architecture we have outlined is *integration*, which has two aspects: firstly, the integration of information relating to components of the product to create a composite picture, and secondly, the integration of the different forms of representation and modelling used. The first is the role of the diagram generator and text generator system components. The use of static 'canned' information, however well-organised and presented, does not fully meet the demands of a flexible production process. The problem is, essentially, that the description of the product as a whole is more than a simple concatenation of the descriptions of its components. Dynamic explanation generation [13] can be used to reduce repetition of information and to produce a more coherent overall product description. This technique, however, applies only to text. A greater challenge is created by the need to pull together all the elements of the documentation of components, textual and other, into a single, integrated product description. This process may be seen as an aspect of product design, and appropriate design techniques, such as the use of CAD software, may have a role here.

The second major function of the system is to assist in the diagnosis of faults in service, and this is the principal purpose of the Model-Based, Rule-Based and Case-Based Reasoning components of the system. Operational simulation via a product model is particularly significant in this respect. Added intelligence may be provided by model-based reasoning [8], in which the model is manipulated by appropriate reasoning routines. A common approach is to use a 'perfect' product model to generate a simulation of the correct behaviour of the system, and then to introduce faults to the model in an attempt to replicate the observed behaviour of the (faulty) real product. Model-based systems have in recent years found favour with industry, for example in the Tiger project [6]. Used in isolation, however, they suffer from the drawback that little or no explanation may be provided to supplement the observed model behaviour, so again, integration with the representational elements of the system is called for. Another goal is to integrate Knowledge-Based Systems with modelling techniques, including model-based reasoning and VR interfaces. Intelligent VR agents (IVRA) [2] adds intelligence to virtual reality to produce autonomous agents that model physical behaviour, using intelligence to guide the use of VR.

The Case-Based Reasoning element is also primarily to support fault diagnosis. CBR systems need not incorporate large databases; very often, a case base of some 100 previous histories will be sufficient to give useful pointers. In the present context, there are issues to be resolved concerning the matching of case data. Because the product under consideration may, in the extreme case, be unique, there may be no case histories relating to this particular model. Hence, it will be necessary for the CBR system to look for matches not just with the circumstances of the present case but also with the product model, in order to find, for example, cases in which a similar but not identical product has been involved.

With respect to the mode of operation illustrated in Figure 2, a further significant feature of the architecture is that of extensibility. New CB, MB and RB may be added as they become available and/or desirable. Similarly, obsolete information providers may be removed without adversely affecting the operation of the system as a whole.

6 Conclusion

The Stoves project attempts to make use of many techniques to provide technical information and diagnostic support to field service engineers. In this paper, we have outlined both an overall system architecture and underlying data structures to provide this function. The key research issues identified centre round the need to integrate the various techniques proposed, and to integrate also information relating to components of the product model to provide a comprehensive overall picture. The next phase of the work will focus on an implementation of the framework described, for which a multi-agent, object-oriented approach is envisaged.

Acknowledgements

The project is supported by the DTI under the Foresight 'Link' programme, and is being carried out by The University of Liverpool in collaboration with Stoves PLC and other industrial partners. The authors wish to thank all those others involved in the project, especially: Stephen Gilmore of EuroServ; Mike Delves of NA Software Ltd; Allan Smith, of the DTI; Stan Price, the project monitor; and Ian Finch from the Connect Centre at the University of Liverpool.

References

1. Aylett, R., Petley, P., Chung, P.W.H., Soutter., J. and Rushton, A (1999). Generating Operating Procedures for Chemical Process Plants. Integrated Manufacturing Systems: The International Journal of Manufacturing Technology Management 10(6), 328-42

2. Ballin, D. and Aylett, R. (1999). Intelligent Virtual Environment and Virtual Agents. Tutorial at 19th SGES In Conf on Knowledge-Based Systems and Applied Artificial Intellgence.

3. Buchanan, B.G. and Shortliffe, E.H. (1984). Rule Based Expert Systems - The MYCIN Experiments of the Stanford Heuristic Programming Project. Addison-Wesley.

4. Carey, R., Bell, G. and Marrin, C. (1997). The Virtual Reality Modeling Language (VRML97). ISO/IEC 14772-1. The VRML Consortium Inc.

5. Crowder, R.M., Wills, G., Heath I. and Hall, W. (1997). The Application of Hypermedia in the Factory Information Environment. Proc 'FACTORY 2000' Conference, Cambridge; IEE Conference Publication 435

6. Edwards, J. 3-Tier Client/Server at Work. Wiley.

7. Falkenhainer, B. and Forbus, K. (1991). Compositional modeling: finding the right model for the job. Artificial Intelligence 51, 1-3

8. Hunt, J.E. (1996). Model Based and Qualitative Reasoning: Industrial Application. Proc 1st Int Workshop on Model-Based and Qualitative Reasoning - Perspectives for Industrial Applications. ECAI'96, Budapest.

9. Hunt, J.E, Lee, M.H. and Price, C.J. (1998). Progress in applying model based and qualitative reasoning to industrial applications. The MONET Newsletter, 1(1).

10. Iwasaki, Y., Farquar, A., Fikes, R. and Rice, J. (1997). A Web-based compositional modelling system for sharing of physical knowledge. Proc 15th Int Joint Conf on AI, AAAI press.

11. Milne, R., Nicol, C., Trave-Massuyes, L., Quevedo, J., Ghallab, M., Bousson, K., Dousson, C., Aguilar, J. and Guasch, A. (1994). TIGER: Real Time Situation Assessment of Dynamic Systems. Intelligent Systems Engineering Journal, 3(3)

12. Moulton, M. (1997). A Rule-Based incident tracking system. In Macintosh, A. and Milne, R. (eds), Applications and Innovations in Expert Systems V, SGES publications, London.

13. Rubinoff, R (1986). Adapting Mumble: Experiences with Natural Language Generation. Proc AAAI-86.

14. Sowiztal, H., Rushworth, K. and Deening, M. (1998). The Java 3D API Specification. Addison-Wesley.

15. Watson, I. (1997). Applying Case-Based Reasoning: techniques for enterprise systems. Morgan Kaufman, London.

16. M. Wooldridge, N. R. Jennings, and D. Kinny. A Methodology for Agent-Oriented Analysis and Design. In O. Etzioni, J. P. Muller, and J. Bradshaw, editors: Agents '99: Proceedings of the Third International Conference on Autonomous Agents Seattle, WA, May 1998.

17. Xue, D., Yadav, S. and Norrie, D.H. Development of an intelligent system for building product design and manufacturing - Part1: product modelling and design.

The Application of Artificial Neural Networks to Anticipate the Average Journey Time of Traffic in the Vicinity of Merges

Mehdi Fallah-Tafti, BSc MSc PhD CEng

Cardiff School of Engineering, The University of Wales, Cardiff, U.K.

SCEMF1@CF.AC.UK

Abstract: A microscopic simulation model representing traffic behaviour in the vicinity of merges, especially under congested situations, was developed. The simulation model was applied to produce a set of data representing traffic patterns in the merge area, ramp metering rates, and the corresponding vehicle journey times. The data were used to develop an artificial neural network (ANN) model, which anticipates the average journey time of mainline vehicles that enter an upstream section during a 30s interval. The ANN model was validated with an independent data set. An investigation was made to ensure that the ANN model and the simulation model are capable of demonstrating the onset of flow breakdown at high combinations of the mainline and the entry ramp traffic flow. The ANN model can be applied to develop an ANN based feedback control system, which adjusts ramp metering rates to keep the average journey times of vehicles close to their desired or target value, and to reduce congestion in the vicinity of merges.

1. Introduction

One of the shortcomings of the current ramp metering strategies, even the robust and successful strategies such as ALINEA [1 & 2] is that they only rely on the local downstream traffic parameters such as occupancy or density as the measure of congestion. Occupancy or density used in the current feedback control systems are local parameters and, when used as the measure of congestion, will not be detected until congestion has reached the measurement section. In addition, these measures cannot represent the traffic pattern, in a long section of motorway, which contributes to the formation of congestion. This causes that these strategies have limited success in preventing the onset of congestion or eliminating that, which occurs.

A potential improvement in the ramp metering strategies then could follow from using those traffic parameters, which represent traffic patterns for a longer section of the merge area. For instance, the average journey time of vehicles in a section from upstream of the merge extending to downstream of the merge is one criterion, which could be used to represent the traffic situation in the merge area. With the current technology, the on site measurement of this parameter in the vicinity of merges for its direct application in the control systems is not feasible. But it could be indirectly estimated by the development of a relationship between this parameter and the measurable traffic parameters such as occupancy, speed, and flow, which can represent traffic pattern in the merge area.

Using the future predicted values of traffic parameters instead of using their current values could also be helpful in predicting congestion. By anticipating potential congestion and reacting appropriately, the performance of a ramp metering strategy would be enhanced. Artificial neural networks (ANNs), have shown promising results in their applications to solve a variety of problems in different fields including prediction type problems in the field of traffic engineering [3, 4, 5, 6 & 7].

This paper describes the process, which led to the development of an ANN model, which uses the traffic parameters of the motorway and the entry ramp to anticipate the average journey time of vehicles in the merge area, using the NeuralWorks Predict software [8].

A microscopic simulation model, which is described in the next section, was applied to provide the required data sets to train and validate the ANN model. The idea was that a simulation model with the ability to represent real traffic behaviour could be used as an alternative to on site traffic studies.

2. The Simulation Model Development

A microscopic simulation model representing traffic behaviour in the normal sections and in the merge sections was developed [9]. A combination of event scanning and time scanning techniques were used to update the decisions of drivers, and the speed and position of vehicles. Graphics and animation were used to enable the user to visually follow the interactions of vehicles during the running of the simulation model (e.g. see Figure 1). A range of input variables was defined in the model, which gives the user the flexibility to simulate traffic flow under a variety of conditions.

The variation in the characteristics of vehicles was considered in the model and these characteristics were randomly assigned to different vehicles. The variation in the characteristics of drivers, including aggressiveness, desired speed, reaction time, etc was also considered in the model. These characteristics were related to each other using the same random number to generate different characteristics of each driver. It was assumed that the reaction time of each driver varies in different situations and also that the reaction time varies across the population of drivers.

New car following, lane changing, and merging algorithms, with emphasise on better representation of the interactions of the mainline traffic and the entry ramp traffic in the merge area, were developed and deployed in the model. The simulation model was calibrated and validated using the site data collected in this study and in previous studies.

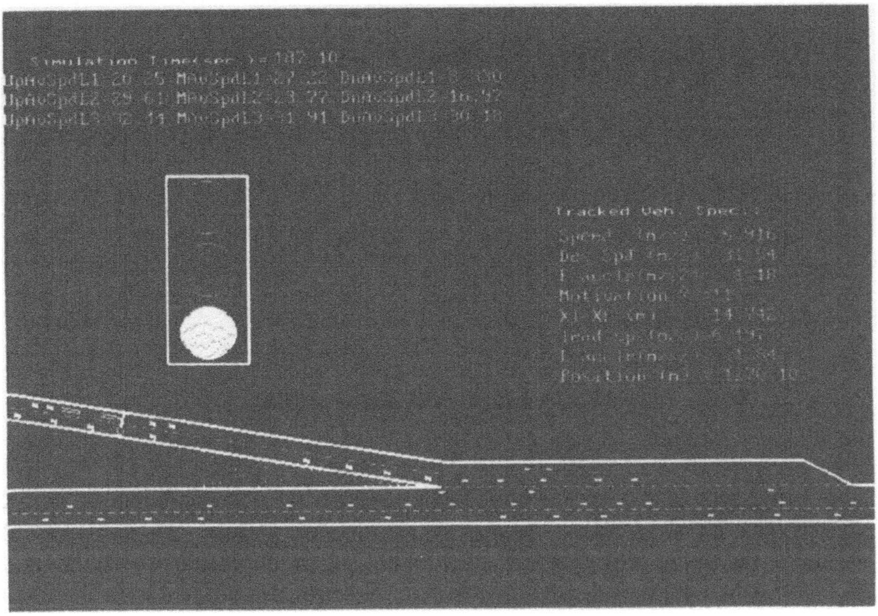

Figure 1 A screen shot of the simulation model

3. The Neural Network Model Development

The Microsoft NeuralWorks Predict [8] was used for the development of the ANN model. This software will be called “Predict” hereafter. This section explains the procedure which led to the development of the journey time predict model using “Predict”.

3.1 Problem Definition

The average journey time of the mainline vehicles that enter an upstream section, shown in Figure 2, along the distance ‘L’ and during a 30s interval, was considered as the output of the ANN model. The interval of 30s was short enough to reflect the variation in the traffic pattern and long enough to produce representative values. This output is assumed to be related to the traffic pattern in the merge area, which could be specified using the relevant traffic parameters. The distance ‘L’ was assumed to be 2.1 km, which is long enough to reflect any breakdown in flow or speed that may initiate downstream of the merge [10].

The parameters average speed, flow, occupancy, and proportion of HGVs are indicative of the traffic situation on the mainline and therefore were considered as the potential input parameters, in the ANN model development. In addition, the metering rate was also considered as one potential input parameter to reflect the effect of merging traffic.

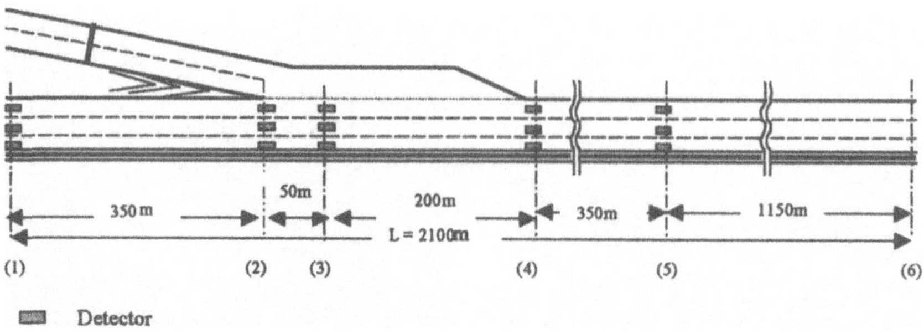

Figure 2 Layout of merge section

The merging traffic changes the traffic pattern on the mainline. It would consequently be reasonable to assume that the average journey time of vehicles is related to the initial traffic pattern measured at a location upstream of the merge, and the traffic pattern measured at a location downstream of the merge. Hence it was decided to investigate the importance of traffic parameters in these two locations as the input parameters of the ANN model. The best positions for the upstream and downstream measurement locations were also investigated, as discussed in the following sections.

Because of the dynamics of traffic flow especially in the merge area, it was expected that the data set produced by the simulation model would be noisy. Efficient model development using noisy data requires minimising the number of input data [8]. Hence it was decided to use the overall average values of traffic parameters on the mainline and on the entry ramp instead of using values for each lane. In previous studies, the occupancy of the inside lane upstream of the merge, and the average speed of vehicles in the outside lane of the mainline at the merge point, have also been considered as the potential measures which can detect the onset of congestion and flow breakdown [11 & 12]. These two parameters were therefore added to the list of the potential input data, which should be examined.

The list of potential input data is outlined below:

- the average speed, flow, and occupancy of vehicles measured at an upstream detector location during the last 30s interval (Up_Spd, Up_Flow, Up_Ocup),
- the average speed, flow, and occupancy of vehicles measured at a downstream detector location during the last 30s interval, (Dn_Spd, Dn_Flow, Dn_Ocup),
- the occupancy of inside lane, measured at an upstream detector location during the last 30s interval (UpL1_Ocup),
- the average speed of outside lane vehicles measured at a downstream detector location during the last 30s interval (DnL3_Spd),
- the percentage of HGVs, measured at a downstream detector location during the last 30s interval (PHGVs), and

➢ the metering rate during the last 30s interval (MTR).

The problem, which is to be solved using the ANN technique, can be expressed in the following mathematical form:

$$Av\_Journt = f(Up\_Spd, Up\_Flow, Up\_Ocup, Dn\_Spd, Dn\_Flow, Dn\_Ocup, UpL1\_Ocup, DnL3\_Spd, PHGVS, MTR) \quad (1)$$

3.2 Preparation of Required Data Examples

The data set, which is to be used to train the neural network model, should include a wide range of relevant traffic conditions. To provide the required data examples, the simulation model was run for a typical merge layout (Figure 2), and a range of combinations of traffic flow, as found in practice for congested situations. The traffic conditions for which the simulation model was run are outlined below:

- Maximum total demand (mainline upstream + entry ramp): 5700, 6000, and 6300 (veh/hr).
- Percentage of the entry ramp traffic: 15, 25, 35 (%).
- Percentage of HGVs: 15,20 (%).

The following process was applied to ensure that a representative range of values was available for the development of the ANN model.

1. The simulation model was set to operate under a 'variable time' metering system in which the cycle period was assumed to be 30s, the maximum green time was assumed to be 24s, the amber time was assumed to be 3s, and the minimum red time was assumed to be 3s. A cycle time of 30s is consistent with most of the traffic responsive strategies applied in this study for comparison.
2. The metering rate in the first cycle of the different runs of the simulation model was set to a minimum value (240 veh/h or 1 veh/cycle/lane). The metering rate was then incremented at a rate of 240 veh/h until the total flow (mainline upstream + entry ramp) reached its maximum (5700, 6000, or 6300 veh/h).
3. The metering rate was then reduced smoothly in further cycles with the same rate as the increment rate until the metering rate was 240 veh/h.
4. Steps 2 and 3 were repeated until the end of the simulation run time period was reached.
5. The above process was repeated in different runs to provide data for a comprehensive set of relevant mainline and ramp traffic conditions.

In each run of the simulation model 130 data examples were generated: repetition for the chosen data input combinations provided 2340 data examples.

3.3 Selecting Train, Test, and Validation Data Sets

The training set is the set of data that are used to fit the parameters of the model. The test set measures how well the model interpolates. It is used as part of the model building process to prevent over-fitting. The validation set is used to estimate model performance in a deployed environment.

In the model development process, two independent data sets, each including 2340 examples and in the manner explained in the previous section, were provided. From the first provided data set, 70 % of data examples were considered as the training set, and the remaining (30%) considered as the test set. "Predict" was set to select randomly these data sets from the first data set. The second provided data set was considered as the validation data set.

3.4 Decision on the Correct Location of 'Detectors'

The neural network model was intended for application as part of a ramp metering strategy. It was essential to select traffic patterns at those sections, which have reasonably high interaction with the oncoming traffic from the entry ramp.

The distance of upstream detectors from the nose was selected so that the average travel time of the motorway vehicles from the detector location to the merge location is almost the same as the required travel time of the entry vehicles from the stop line to the acceleration lane. Using this detector position, the mainline and entry ramp traffic, for which the parameters have been measured in the same cycle, would be more likely to have a direct interaction in the merge area. Consequently, the average journey time of those mainline vehicles, which start their journey from the upstream detectors during a cycle, would be more likely to be affected by the introduced metering rate for that cycle. On this basis and through several runs of the simulation model especially around the congestion situations, the location of upstream detectors was fixed at 350m from the tip of the nose.

The average journey time of vehicles in the merge area could also be affected by the traffic pattern downstream of the merge. Four different locations were therefore considered as alternative locations for the downstream detectors. These locations were at the nose, 50m downstream of the nose, at the end of the acceleration lane, and 600m from the nose (see Figure 2). During the simulation model runs, the required downstream traffic parameters, described in section 3.1, were recorded at each of the above locations. The results of the simulation model were divided into four different sets each corresponding to one downstream measurement location but with similar upstream and entry ramp traffic parameters. Each data set was used to develop an ANN model which predicts the average journey time of vehicles in each cycle according to their upstream, downstream, and entry ramp traffic parameters given as input. The performance of the ANN models was then examined and compared using different statistical measures.

The model in which the downstream measurements were obtained from the end of acceleration lane demonstrated the best performance. The best position for the downstream detector location was therefore decided as the end of acceleration lane and further investigation was concentrated on the data provided from this location. The range of variation of data, which were used for the development of this model, is shown in Figure 3. As indicated in this Figure, a sufficient range of variation for each input parameter has been produced in the data set.

Among introduced input variables to the ANN models, UPL1_Ocup and DnL3_Spd were rejected as input variables in the variable selection stage of all of the developed ANN models. These two input variables were therefore eliminated as potential input variables in the further analysis.

3.5 Investigation of the Possible Further Improvement of the Developed ANN Model

In the previous section, the default settings of "Predict" were used to develop the ANN models. This mode requires a minimum level of intervention by the user. The default settings are usually adequate for a wide variety of problems. However, in some problems, an adjustment in one or more of the default settings may lead to a model with an overall better performance. In some cases a less complex solution, for example a reduced number of PEs in the hidden layer, may lead to only a marginal reduction in performance.

The possibility of further enhancement of the selected ANN model, developed in the previous section, in the aspects of its performance and network size was investigated through the development of several new ANN models in which the default settings of "Predict" were individually adjusted. These models are outlined below.

- A new ANN model was developed in which no transformation of input variables was used.

- A new ANN model was developed in which the data was assumed noisy and hereby the Kalman learning rule, which is more suitable for the noisy data, was applied for the model training.

- Two new ANN models were developed in which a lower limit for the maximum number of PEs in the hidden layer was imposed. In the model developed using the default settings of "Predict", the maximum number of PEs in the hidden layer was restricted to 30, and consequently, 8 PEs were obtained for the hidden layer. In the development of two new ANN models, the upper limit for the number of PEs in the hidden layer was set to be 10 and 6 PEs.

- An attempt was made to see if any further input variables could be removed to reduce the network size without substantially reducing model performance. The percentage of heavy goods vehicles is more difficult than traffic flow to measure automatically on site and was omitted from the input variables to assess the effect.

The characteristics and performance of the new developed ANN models after each of the above alterations were then compared. The following criteria, which are sometimes conflicting, were used to compare the developed models and select the best model for this study.

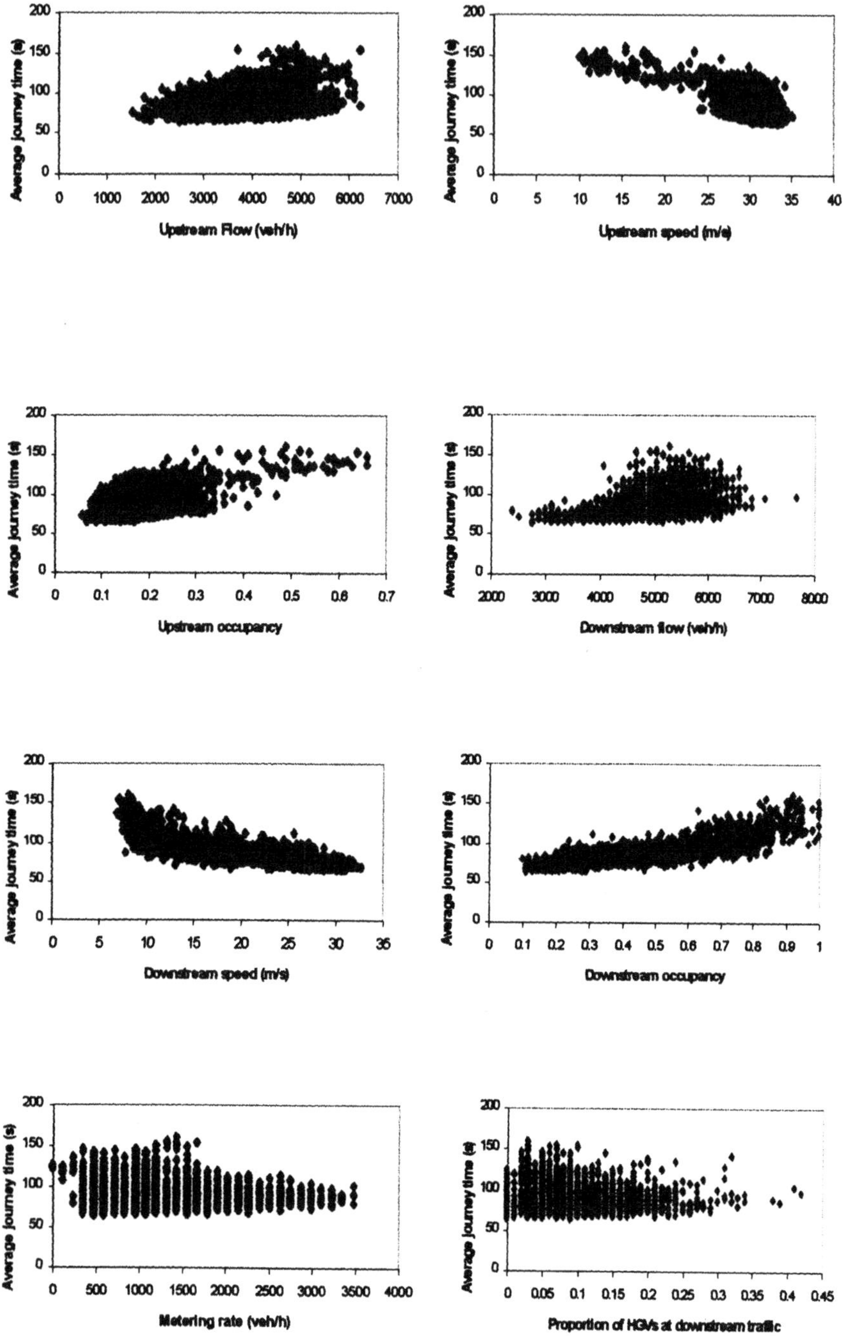

Figure 3 The range of variation of data used to develop the ANN model

- the performance of the selected model, based on the statistical measures, should be better than the other models, and
- from the practical application point of view and economic aspects, a model with the minimum required input data, data transformations, and hidden layer size should be chosen.

On this basis, the model, which was developed using non-transformed input data, was selected as the best model.

3.6 Evaluation of the Developed ANN Model

Initial results of the measures of performance of the selected ANN model indicated that the performance of the model for the training, test, and validation data sets was similar. This indicated that the generalisation ability of the developed model was satisfactory. However, the effectiveness of the developed ANN model can only really be measured by its comparison with other models on previously unseen data.

Non-linear feedforward ANNs are a generalisation of the more common techniques of linear and logistic regression and they should presumably provide more accurate results. The Multiple Linear Regression (MLR) technique was therefore used as an alternative method to assess the performance of the ANN model. The following MLR models were developed, using different data sets which were obtained from the first provided data set (train set + test set), and their performance was compared to the developed ANN model:

- a MLR model was developed using the provided data examples and all of the input variables, and
- a MLR model was developed using the provided data examples and those input variables which were used in the selected ANN model.

The performance of the developed MLR models in comparison with the selected ANN model is demonstrated in Table 1. As indicated in this Table, even though the MLR models demonstrated similar performance on the (Train + Test) data set but they did not show a good performance on the validation data set. Thus it can be concluded that the developed ANN model is more effective than the MLR models for this application purpose.

A brief definition of measures used for comparison follows:

- **R:** the Pearson correlation between the real target outputs and the model outputs.
- **Avg. abs. error:** the average absolute error between the target outputs and the model outputs.
- **RMS:** the root mean square error between the target outputs and the model outputs.

Table 1 Comparison of the performance of the ANN model with the alternative Multiple Linear Regression models

Model type	Data set	R	Av. abs. error (s)	RMS (s)	No. of inputs
Finally selected ANN model	Train +Test	0.93	4.44	5.94	6
	Validation	0.92	4.70	6.23	
MLR model developed using all input variables	Train +Test	0.92	4.70	6.19	8
	Validation	0.62	6.00	16.44	
MLR model developed using input variables of the selected ANN model	Train +Test	0.92	4.78	6.27	6
	Validation	0.62	5.99	16.01	

3.7 Validation of the Developed ANN Model

3.7.1 Comparison of the ANN model results under the validation data set with the corresponding results from the simulation model

For the validation of the ANN model, the scatter diagram of the model outputs and their corresponding real outputs was provided and the line y = x was also imposed for the guidance (see Figure 4).

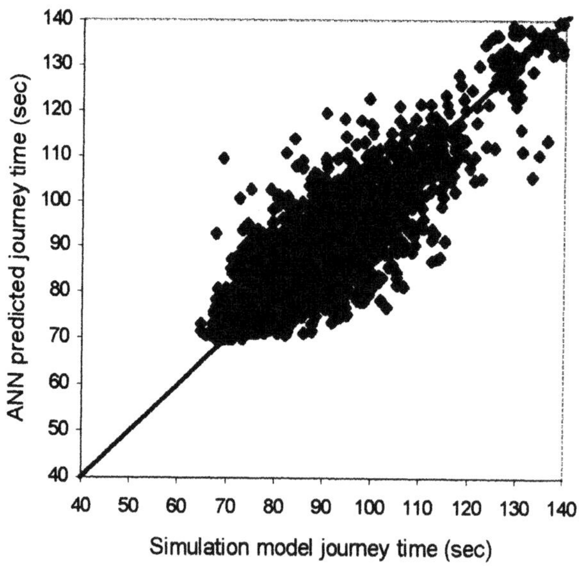

Figure 4 Comparison of the simulation journey times in the validation data set with the corresponding values predicted by the ANN model

As graphically demonstrated in this Figure, the data are scattered around the line y = x, which indicates that there is no bias, with a good correlation between the ANN model outputs and the simulation model outputs (R=0.92). The performance of the model is therefore satisfactory.

3.7.2 Comparison of the results of the ANN model at different entry ramp traffic flows with the corresponding values from the simulation model

The online performance of the ANN model at different entry ramp traffic flows was investigated to ensure that the ANN model is capable of producing the same results as the simulation model, and that both models are capable of demonstrating the occurrence of flow breakdown at high combinations of the mainline and the entry ramp traffic flow.

The simulation model was set to operate under a constant upstream traffic flow but with a variable traffic flow from the entry ramp in different runs. The entry ramp traffic flow was incremented in consecutive runs from a minimum of 240 veh/h to a point at which the further increase did not lead to the practical increase in the entry ramp traffic flow (because of the capacity limit of the merge layout). For each entry flow rate, the simulation model was run several times and the same random number seeds were used for different entry ramp traffic flows. For each entry ramp traffic flow rate, the average journey time of mainline vehicles along the distance 'L', shown in Figure 2, was measured and compared with the corresponding values from the ANN model. Figure 5 compares the results of the simulation model with the ANN model at different entry ramp traffic flows at three constant upstream traffic flows (3000, 4000, and 5000 veh/h).

As indicated in this Figure, there is a close proximity between the results of the ANN model and the simulation model under similar conditions and the performance of the ANN model is therefore satisfactory. The results confirm that both models are capable of demonstrating the onset of flow breakdown, based on increased journey time, at high combinations of the mainline and the entry ramp traffic flow.

3.8 The Structure of the Finally Selected ANN Model

The structure of the finally selected ANN model, including the model inputs, and the model output and the corresponding transform function are shown in Figure 6. As indicated in this Figure, the ANN structure has one hidden layer, which consists of two PEs. There is no transformation of input data. The input data to the ANN model can be directly obtained from the onsite traffic detectors.

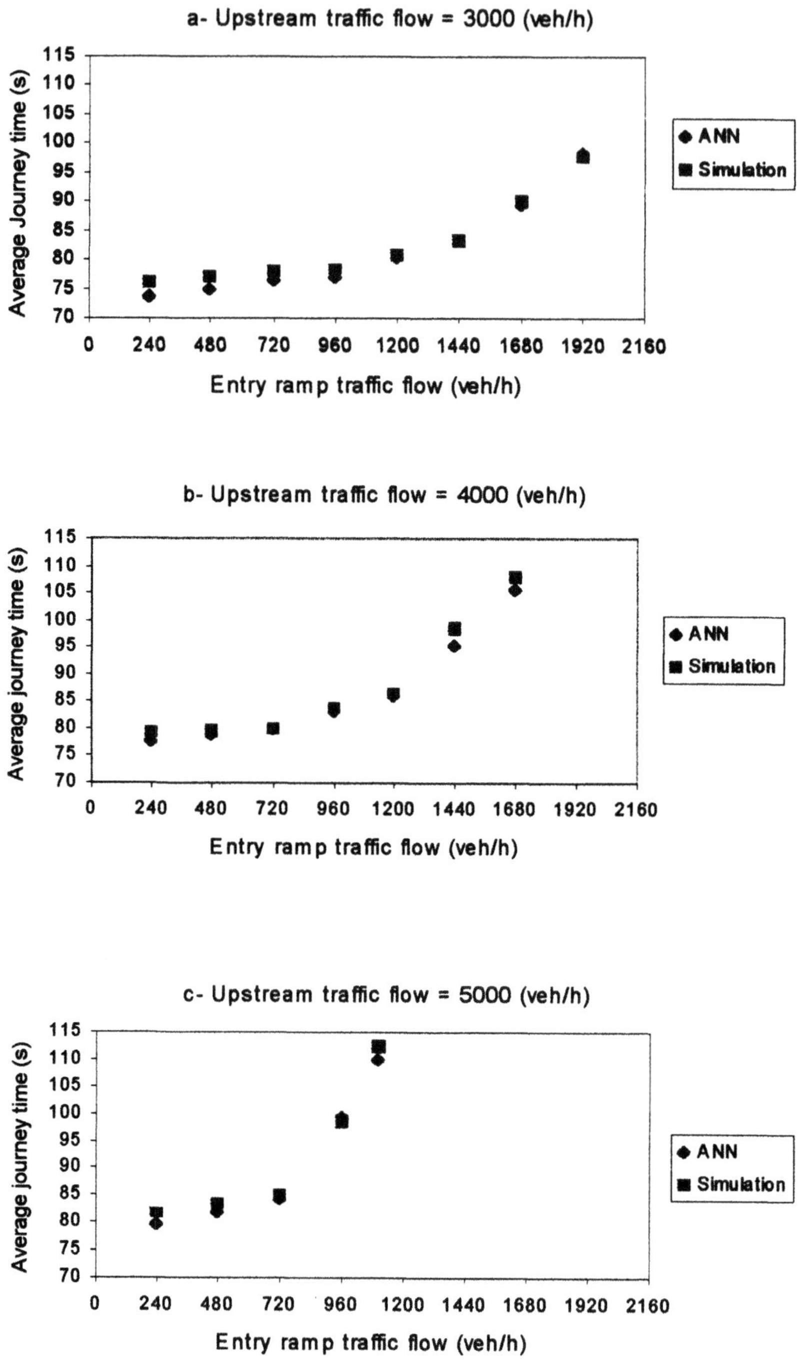

Figure 5 Comparison of the variation of the average journey time of vehicles with the entry ramp traffic flow obtained from the ANN model and the simulation model

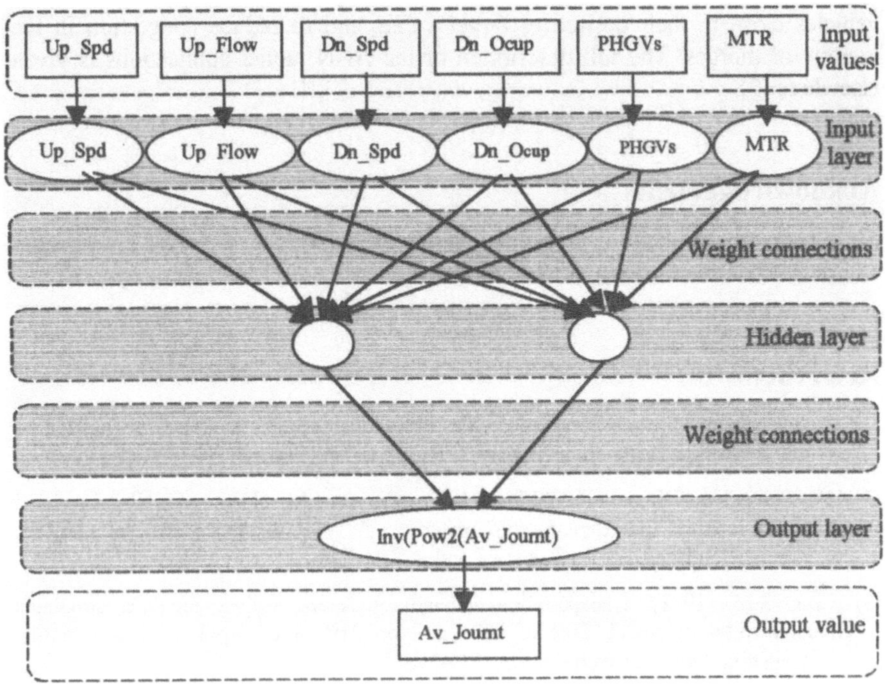

Figure 6 The Structure of the developed ANN model

4 Summary and Conclusions

A microscopic simulation model was applied to produce a set of data representing traffic patterns in the merge area, ramp metering rates, and the corresponding vehicle journey times. The data were used to develop several ANN models, which anticipate the average journey time of mainline vehicles that enter an upstream section during a 30s interval. The mainline traffic parameters at a merge upstream location and a merge downstream location and also the metering rate, all measured during the last 30s interval, were used as inputs to the ANN models. The developed ANN models were then compared. The ANN model in which the non-transformed input data was used, demonstrated the best performance. The ANN model was validated with an independent data set. Statistical analysis and validation processes confirmed that the generalisation ability and the accuracy of the selected ANN model are both satisfactory. The selected ANN model was also compared with the alternative Multiple Linear Regression models. The results confirmed that the ANN model has an overall better performance. A comparison was made to ensure that there is a close proximity between the results of the simulation model and the ANN model and that both of the models are capable of demonstrating the occurrence of flow breakdown. The outcome of this investigation was promising. The ANN model could be applied to develop an ANN based feedback control system, which adjusts ramp metering rates to keep the average journey times of vehicles close to their desired or target value, and to reduce congestion in the vicinity of merges. The full description of the ANN model applications is given elsewhere [9].

Acknowledgements

The author would like to express his gratitude to Dr John G. Hunt for his valuable advice and comments during this study.

References

[1] Papageorgiou, M., and Blosseville, J.M. (1989). Macroscopic modeling of traffic flow on the Boulevard Peripherique in Paris. Transportation Research B 23, 29-47.

[2] Haj-Salem, H., Blosselville, J.M., and Papageorgiou, M. (1990). ALINEA- A local feedback control law for on-ramp metering: A real life study. 3rd International Conference on Road Traffic Control, London, 194-198.

[3] Chin, S.M. (1992). Transportation demand forecasting with a computer simulated Neural Network model. Proc of Int. Conf. on Artf. Intel. Appl. in Transportation Engineering, San Buenaventura, CA, 349-390.

[4] Dougherty, M.S., Kirby, H.R., and Boyle, R.D. (1993). The use of Neural Networks to recognize and predict traffic congestion. Traffic Engineering and Control 34(6), 311-314.

[5] Chang, G.L., and Su, C.C., (1995). Predicting intersection queue with neural network models. Transportation Research C 13(3), 175-191.

[6] Ledoux, C. (1996). An urban traffic control system integrating Neural Networks. IEE Conf. on Road Traffic Monitoring and Control, Conf. Publ. No. 422.

[7] Zhang, H.M., and Ritchie, S.G. (1997). Freeway ramp metering using Artificial Neural Networks. Transportation Research C 5, 273-286.

[8] NeuralWare Ltd. (1995). NeuralWorks Predict: Complete solution for neural data modeling. NeuralWare, Inc., Pittsburgh.

[9] Fallah-Tafti, M. (1999). The application of simulation and Neural Networks to develop a ramp metering strategy, PhD Dissertation, The University of Wales, Cardiff, UK.

[10] Hounsell, N.B., Barnard, S.R., and McDonald, M. (1992). An investigation of flow breakdown and merge capacity on motorways. Transport Research Laboratory, Berkshire.

[11] Owens, D., and Schofield, M.J. (1988). Access control on the M6 motorway: Evaluation of Britain's first ramp metering scheme. Traffic Engineering and Control 29 (12), 616-623.

[12] Taylor, C, and Meldrum, D. (1995). Simulation testing of a Fuzzy Neural Ramp metering algorithm. Washington State Department of Transportation, Olympia, Washington.

A KBS Framework for Scheduling Structural Maintenance

S.M.C. Peers
University College London,
s.peers@meng.ucl.ac.uk
http://www.meng.ucl.ac.uk/

Abstract: Inspections, repairs and maintenance (IRM) actions are carried out periodically on many large structures, such as fixed offshore structures and bridges, to ensure the integrity of the structure. For fixed jacket platforms in the North Sea, the results of reliability-based fatigue fracture mechanics analysis can be used to target the most likely components requiring at least inspection and possibly repair. However as there are considerable costs associated with inspection and maintenance of platforms, it is necessary to optimise the IRM plans whilst keeping risk within acceptable limits. Further, it is a practical and absolute requirement that all plans conform to any guidelines set out by the governing authorities.

The planning problem is to identify possible IRM actions. The scheduling problem is to combine actions into one schedule ensuring that regulations are not infringed. This paper describes a framework that uses a generic search strategy including constraint satisfaction. A KBS can provide support in planning IRM tasks for fixed offshore platforms with a reliability-based fatigue fracture mechanics package. The paper also briefly reviews the use of AI in the offshore industry and for planning and scheduling.

1. Introduction

Offshore jacket type structures are highly redundant structures subject to many uncertainties and in particular, to time-varying, cyclic loads and so to fatigue. For many reasons, such as the degradation of existing structures, there is a need to be able to analyse the reliability of these structures in a rigorous manner. Hence reliability techniques developed specifically for complex structures are used.

The real problem lies in applying reliability analysis to the production of rational inspection strategies. To employ reliability analysis for this application requires understanding the inspection procedure in order to choose the appropriate information on inspection techniques. There are also issues in generating input data for the analysis and in interpreting the output. The interpreted results must then be combined to form a rational global solution for the structure that takes into account constraints that may not have been included, implicitly or explicitly, into the analysis. It is necessary to develop complete systems that will aid in carrying out the analysis and interpreting the results.

One approach adopted was to provide a computer-based tool that would estimate the expected total costs associated with a particular inspection and maintenance plan. In that way, different inspection and maintenance plans can be compared, constraints enforced and from this an appropriate rational plan chosen.

This paper briefly reviews AI applications in the offshore industry and AI-based scheduling before describing a knowledge-based reliability IRM planning and scheduling system.

1.1 Background to Reliability-based IRM Planning

Since the mid-1980s, reliability methods have become accepted tools in several industries such as offshore, nuclear and aerospace. The potential of these methods as decision support tools in engineering lies in the consistent modelling of uncertainties that enables comparisons of different actions and designs. In more recent years, modern reliability methods within classical decision theory applied to inspection, repair and maintenance (IRM) planning of engineering systems subject to uncertain deterioration processes has been investigated (see [1]). The work has led to a consistent framework for evaluation, as expected costs, of different inspection and repair actions. Cost evaluation allows optimisation of the overall inspection and maintenance plan for a given structure within its lifetime.

The reliability of a structure may be maintained by carrying out repair operations at various scales and more frequent inspections. The costs and benefits of the options could be assessed in part by making use of reliability-based fracture mechanics analysis. Fatigue fracture mechanics theory provides the mechanism for predicting the state of the structure for the purpose of planning future IRM actions. Structural reliability analysis provides decision-makers with objective measures by which to base their IRM planning policies. This provides the background to a Reliability-based Inspection SCheduling (RISC) methodology for the sub-deck tubular welded connections of jacket-type platforms (see background detail in [2, 3]).

There are, however, difficulties in applying structural reliability analysis related to the amount of input data, interpretation of the results and so on, as these techniques are relatively new in the field. These techniques require many different forms of data from many different sources and each operator or application may require a different interpretation. There is also the problem of somehow combining individual inspection actions to provide a rational schedule for the whole structure taking into account the constraints on operators. Furthermore, as for any computer controlled decision process, there are the implicit requirements of

- flexibility of the representation scheme for the different types of information
- effective handling of incomplete information on the structure
- selective utilisation of knowledge for efficiency
- the capability for operators to control the problem solving sequence
- ease of maintenance and extension as new information is gathered and as procedures change

To implement a reliability-based IRM scheduling methodology, a knowledge base system is indicated that carries out the required analysis and will aid planning.

2. AI-based Decision Support

Scheduling and planning are decision processes, which involve selecting the most appropriate sequence or set of actions to be carried in the future, given an initial state. Yang [4] distinguishes carefully between the two terms. Planning is the production of a description of how events are interrelated and how to sequence them, which would include the definition of the constraints on the sequence of the events. The aim of scheduling, on the other hand, is to select an optimal set of events, which will satisfy certain conditions or constraints.

Scheduling and planning are real-world problems to which mathematical techniques borrowed from operational research and management science have been applied, such as linear programming for optimisation of manufacturing plans and dynamic programming for assembly scheduling. Although these methods are often effective in certain situations, for the general scheduling problem they can fail [5]. AI has been applied to this area in an attempt to take into account all the problems that arise in scheduling which cannot be taken into account by the use of mathematical techniques.

Furthermore AI concepts have been applied extensively to the whole range of real world engineering problems and in particular to offshore applications in an effort to model the deep knowledge employed by humans.

2.1 Offshore Applications

The oil industry very rapidly grew interested in expert systems during the 1980s following the early and very successful Prospector system [6]. These early engineering applications of KBS were based on the production rule system architecture. An example is Gasoil, which provided advice at BP International on many engineering tasks on petrochemical rigs [7]. Gasoil was reported to be the second largest expert system of its time in that it has a knowledge base of 2500 rules extracted by induction on examples provided by experts. Another oil-company, Amoco Production Company, provided programs used by well-logging analysts to be incorporated into ELAS (expert log analysis system), which is described as a hybrid system including experiential heuristics with mathematical methods [8]. Other early expert systems for marine applications included: FOCES (fuel oil characterisation expert system) which aided ranking of fuel combustion behaviour; and a diesel engine diagnostic system, DEEDS, which contained a database of symptoms and associated faults, and an engine simulation model [9]. Systems were also developed to optimise maintenance schedules for hull and machinery of ships and have been applied to the use of analysis programs [8,9]. At least one KBS has been developed for the assessment of flaws in welded components employing codes of practice and a fatigue advice system makes use of hypertext within an expert system for use during design [10,11].

Of particular interest is the use of AI techniques in the reliability assessment of offshore structures and for deciding on an inspection strategy. Two systems have been developed in the past which attempt to do this. The RAMINO (Reliability

Assessment for Maintenance and Inspection Optimisation) system aided in carrying out reliability analysis on offshore oil platforms and on pressure vessels of nuclear power plants [12]. RAMINO provided three levels of knowledge, the analysis programs, object-rules and an overview super-module, and aimed to consider many different failure modes. The other system, PLAIM (Platform Lifetime Assessment through Analysis Inspection and Maintenance), provided assessments of fixed jacket platforms and planning for IRM tasks [13]. PLAIM, in common with RAMINO, provided a non-prescriptive approach to the assessment of the reliability, unlike the approach taken for the system described later in this paper. The large rule bases, more than 1000 rules required for 100 tasks, employed in PLAIM would result in problems of knowledge maintenance and in control of the reasoning. A more general decision support system for safety management for the offshore industry is ARMS (Advanced Risk Management System) [14]. This is a risk assessment tool using heuristics and decision theory to build influence diagrams and provide operators with *what-if* analysis to identify areas of concern for any offshore structure.

More directly relevant to structural reliability is work described by Chen et al [15] on using AI-based searching techniques in evaluating structural system reliability. In this, the systems models used in the so-called Beta-unzipping Method for estimating system reliability of structures are re-formulated as search trees. AI-based searching techniques are applied to finding the system probability of failure more rapidly than by the usual method. Also of long-term interest is how to incorporate probabilistic human error analysis: a recent rule-based system HuRA attempts to do just this [16].

2.2 Scheduling and Planning

Planning problems involve ordering the sequence of activities. This requires identifying precedences, that is, any relationships between activities that dictate the order of execution. For instance, if a detailed IRM schedule were required, it would be necessary to realise that inspection of a joint must be carried out before a repair, thus inspection has precedence over repair.

For activities with no direct relationship, it may be possible to run these in parallel. Plans for this situation are termed *non-linear plans* and can be modelled as a network. The planning problem for reliability-based IRM of offshore structures, however, does not require a detailed schedule showing the order of the activities in each period. Thus, network planning will not be considered here.

Scheduling involves allocating resources to perform a series of tasks over a period of time. The decision problem here is to how to assign actions to points in time, to achieve the goal of ensuring that all the actions are carried out within some stated objectives. Le Pape [17] defines two basic categories of scheduling problems:

- pure resource allocation, in which the demand for each resource is known beforehand and the problem is to ensure that resources are allocated such that availability equals or exceeds the demand

- pure scheduling, where the capacity of each resource is known beforehand and the problem is to schedule activities such that the demand on each resource is less than or equal to the availability

Realistic scheduling and planning problems are combinations of the two types.

Researchers in AI are interested in planning on two levels. On the one hand the planning problem occurs in many situations: the human is a planning animal that will formulate detailed programmes of actions. In addition, AI programs require an element of planning to solve a particular problem. In both cases, there is the common aspect of intelligent agents trying to accomplish goals.

Planning and scheduling can be formulated as searching problems, since the aim can be loosely stated as to find a solution out of all the possible actions, such that the required goals are achieved [18,19]. A plan or schedule may also be produced by case-based reasoning [20]. If a past plan was successful then a similar plan is likely to succeed again, as long as the objectives remain the same and there have been few changes to the allowable actions. This reflects reasoning frequently employed: if a plan used for a previous similar situation does not succeed in the current situation, then a slightly modified one will and this can be found by trial-and-error [18, 21].

One way of searching through a range of possible combinations of decisions is to consider the solution space to be a tree of possibilities otherwise known as an OR tree [22]. A node of an OR tree represents one event and the arcs from each node represent alternative events, only one of which can occur. In the tree of possibilities, each node is a decision and from each node, there is another set of possible options from which to choose.

2.2.1 Constraints-Based Planning

Scheduling problems can be dealt with by utilising constraints satisfaction schemes that exploit the declarative representation of the restrictions on resources. Many researchers have made use of constraints-based ideas in scheduling systems for many different industries [23].

A constraint represents a relationship between plan variables as a condition that must be satisfied. So, "*sub-sea inspections of tubular welded connections can only be carried out during the summer months*" is a general constraint representing the notion of a weather window. A more specific constraint may be "*all joints to be inspected with eddy current techniques must be inspected in one weather window*" representing the operator's concern that the specialist divers trained to use eddy current equipment may be employed for one inspection period only. Certain constraints are dealt with implicitly by allowing only a certain possible set of values; others are dealt with by employing heuristics or by systematic methods.

One method of dealing with constraints is to carry out a generation, posting and satisfaction cycle [24]. By *generation,* it is meant that new constraints are formulated given the current state of the plan. As the plan is refined, then the constraints are also refined. During *propagation*, constraints are associated with a network describing the scheduling process. Thus, the constraint "*sub-sea inspections of tubular welded connections can only be carried out during the summer months*" becomes "*only one month remains free for inspection tasks in year 3*" at the point where a specific weather window is under consideration and several inspection tasks have already been allocated. This process ensures efficient searching and provides a method of reducing the search space.

More systematic methods exist for general constraint satisfaction problems (CSP). Kumar [25] and by Bolc and Cytowski [22] describe a CSP as a set of variables with a finite set of discrete values for each variable and constraints defined on the variables. The CSP is then a network where nodes represent variables and arcs between pairs of variables represent constraints. The methods allow systematic searching and reduction of the graph until no solution or one or more solutions are given. These methods treat all constraints in the same way and are restricted to discrete valued variables.

It can be useful to categorise scheduling constraints to allow different treatments. Hard constraints represent physical restrictions or legal requirements, which cannot be exceeded and hence these predominate in the creation of an initial schedule. The latter are preferences and can be exceeded if a solution is otherwise difficult to find. The CSP algorithms described by Kumar target hard constraints. Other researchers have considered ways of modelling soft constraints to allow uncertainties in requirements to be considered; e.g. Berry [6] describes using statistical models of the constraints for a manufacturing application.

3. A KBS for IRM Planning

This section describes the design of a demonstrator system that forms the basis for a complete system for the rational reliability-based IRM scheduling for fixed offshore platforms. Interviews with operators confirmed that the exact decision-making procedure in IRM of platforms is difficult to identify. However, the knowledge and information acquired on analytical tools allowed functional subsystems of the overall system to be identified. These can be summarised as: the evaluation of reliability and costs associated with IRM plans for each of the welded joints of the platform, based on fatigue as a failure criteria and employing past inspection results; finally scheduling for the whole structure taking into account constraints.

The main interactive user is the scheduling and repair engineer in the maintenance department, who has a good understanding of the scheduling process. The two main outputs are the schedule of inspection and repairs, and a detailed log on the analysis results. User interaction involves viewing stored knowledge, incorporating temporary or operator-preferred constraints and heuristics and modifying the knowledge bases permanently to reflect changes to regulations and guidelines that may affect scheduling constraints.

The system needs to be expandable and extensible to allow, for instance, new structural information and updating of guidelines. The reasoning behind derivation of input data to the analysis, the interpretation of analysis results, and the generation of schedules needs to be made explicit to allow users to question results and to interact with the system. These requirements in particular suggest a knowledge base system approach is appropriate.

3.1 The Scheduling Model

The base decision procedure is rooted in structural reliability analysis, which requires much data. Interpreting the analysis results and combining of possible inspection actions into a rational schedule requires many forms of information. The KBS to implement RISC includes external analysis modules and databases integrated with knowledge bases with information on the use of the modules.

The domain sizes of the variables are, in theory and at least initially, large, that is there are many possible IRM actions for each joint or node. The resource constraints on the resources are flexible in that an operator can decide to re-allocate resources from one year to another albeit within certain budgetary limits. The more systematic CSP algorithms were not considered in detail, instead a pragmatic approach involving a searching for possible schedules was taken.

The planning problem is solved by applying heuristics representing hard constraints to narrow down the number of possible actions for the structure. These actions are evaluated by use of reliability analysis and a subset of *best* plans, which may be several for each joint, are then selected and form an initial schedule. This is then modified to take into account the softer constraints.

3.1.1 Outline of Analysis Routes

The overall concept for the complete scheduling procedure is as follows.

1. From the state of the structure a set of joints is selected for initial consideration.
2. According to the current state of the structure and of the joint, including any preferred analysis routes, an analysis route is chosen for the joint. A simplifying assumption is that inspection decisions are independent.
3. The outcome of the analysis is a set of proposed inspection actions, which may also include suggested repair actions.
4. The proposed actions are combined into one schedule for the structure. At this stage it is convenient to consider if any other joints should also be considered for inspection and the user may make changes to the generated schedule.

Each analysis route can be quite complex and may involve shared elements; for example, some rule-based reasoning knowledge sources including regulatory requirements, or the same analysis module but with different inputs. The most usual route however is that of Cost Evaluation of Maintenance Plans which shares the reliability-based fatigue fracture mechanics module (RFFM) with the Reliability-based Evaluation route.

<u>Heuristics-based Evaluation</u> The planning of inspection of joints for which RFFM analysis is inappropriate, which applies for instance to damaged joints, grouted joints, etc. requires rule-based reasoning. The rules here would include simple engineering heuristics as well as those derived from the regulations.

Fatigue Life-based Evaluation It may be possible that operators may wish to consider only a deterministic evaluation of the predicted crack growth of a joint.

Optimal Scheduling For a very limited number of joints, mathematical optimal scheduling could be carried out. This is likely to have a limited influence on the final cost of the schedule and is computationally expensive, hence is not considered any further here.

The RFFM Routes The Reliability-based Evaluation route provides for individual joints or nodes a set of possible inspection times using only reliability measures. The Cost Evaluation of Maintenance Plans route, on the other hand, allows analysis to evaluate expected costs for each plan given proposed possible inspection plans (points in time, technique and repair method).

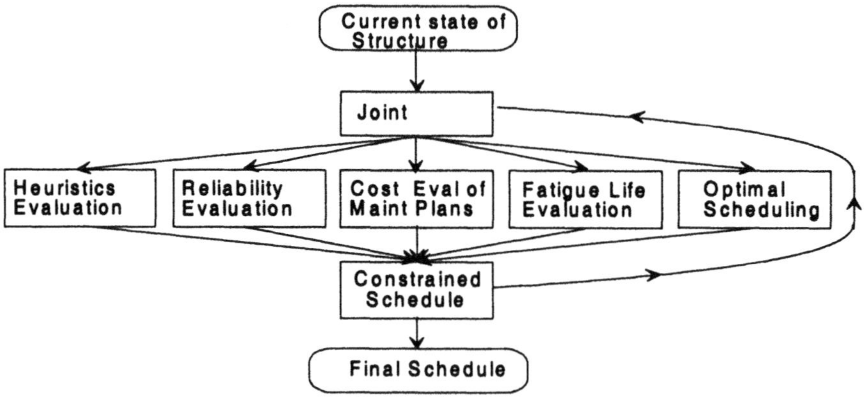

Figure 1: Analysis Routes

Both routes require the RFFM analysis module. This was implemented as a set of linked software sub-modules including a loading pre-analysis to provide the required stress history data; component level Level II structural reliability methods including Bayesian updating; failure criteria based on fatigue fracture mechanics. The inputs for RFFM analysis are

- joint dimensions, the initial crack size and the crack size at failure, with associated uncertainties
- the last inspection result with associated repairs, if any, and uncertainties
- maintenance plans, in the form of a proposed inspection technique with its associated reliability, a repair strategy, and a set of inspection times
- stress concentration factors and the stress history information from the structural model and databases and material properties from a database
- analysis options, stored with the joint information, or as global structural data or as default operator's preferences.
- costs on failure of the component, of the inspection and of repair.

The possible outputs from the RFFM analysis is an initial deterministic assessment of the crack growth, reliability measure (the beta value) over points in time, and the cost evaluation of maintenance plans.

3.2 System Components

The table below summarises the engineering analysis modules (AM), databases (DB) and knowledge base system components (KB) of the system.

Component	Function	Type
System Controller	Overall control of other components	KB
Set-up Procedure	Sets up new structural model with the appropriate loading data and initial reliability analysis results	KB
Analysis Modules and Databases KSs	Expected route and interfaces defined by the operator and expert engineers	KB
Loading Module	Calculates stress concentration factors for a joint	AM
Ranking KS	Selects joints for consideration and cost evaluation analysis	KB
Maintenance Plans	Proposes maintenance plans to be considered for a joint	KB
RFFM Module	Calculates expected costs for maintenance plans and reliability measure for a joint	AM
Results Interpretation	Identifies inspection actions from output of analysis	KB
CS Scheduling	Generates schedules based on Constraints Satisfaction	KB
Observations	Interpretation of inspection results to update structural model	KB
System Manager	Editing functions and maintenance of the system	KB
Structural Model	Structural data is stored as a set of DBs for efficiency	DB
Inspection Methods	Contains inspection reliability data for each technique	DB
Materials	Contains material properties data	DB
SCF	Stores stress concentration factors as a look-up table	DB
User interface	Runs the GUI and models users	KB

Table: The System Components

Of the above components, the databases and analysis modules were completed using standard programming languages and libraries. The KBS elements were completed to a prototype stage: in particular Maintenance Plan and CS Scheduling have been fully demonstrated. More work is needed to make the system usable in practice, such as on the Set-up and Observations modules and the User Interface.

The knowledge bases contain information to be used by the analysis and scheduling procedures. In particular, it is the explicit data and information for the structure in question that will enable rational decisions to be made during the production of a schedule. The Structural Model stores knowledge that can be broadly classified as that arising from design and construction, or the results of analyses, and inspection and environmental data. This KB holds only data used in analysis and scheduling. A knowledge base holding knowledge sources for each external analysis module and database is needed. As each external analysis is called during set-up or at other stages, then the appropriate KS is called to provide information on the format of the input and output (typically text-files), method of execution and error interpretation. If a database is to be accessed then similarly a KS provides the paths and the mapping to the table.

Ranking KS One of the tasks prior to analysis is to create a list of joints to be considered for analysis, as it would be time consuming to analyse all the joints. The procedure does not identify joints that require inspection, instead it provides a priority listing for analysis only. For these ranked joints, the KBS creates input datafiles for the reliability analysis by accessing the Structural Model and gathering all other required data.

A proposed halting condition for the analysis is to stop analysis of joints further down the ranked list when a joint is found to have an optimal inspection time long after the end of the scheduling period of interest. This strategy represents the heuristic that all joints of interest are likely to be above that joint in the ranked list.

Maintenance Plans KS In order to calculate expected costs for various given maintenance plans for a joint, the RFFM module requires definition of the plans as the set of inspection times to be considered, the inspection technique to be used and a repair criteria representing the chosen repair strategy. Inspection techniques and repair criteria are assumed or derived from the information stored on the joint on preferred or allowable options.

The main task of this rule-based KS is to derive a set of inspection times, denoted by t_1 to t_N. To be able to gauge whether the optimum inspection time for a joint is beyond the end of the scheduling period, the possible inspection times must be increased beyond the scheduling period, up to t_{N+5}, say. As safety should not be compromised, i.e. this is a hard constraint, the following rule is applied:

If	the reliability index of the joint falls below its allowed threshold value by inspection period t_n where $n < N+k$
then	the set of possible inspection periods will be t_1 to t_{n-1}
else	the number of possible inspection periods will be t_1 to t_{N+k}.

Figure 2 Inspection times Rule

Results Interpretation This KS provides an initial suggested schedule and suggests alternative actions. Usually, the output from the Cost Evaluation analysis is easily interpreted. The minimum expected cost action is selected initially. Alternative actions will be considered if the minimum expected cost plan in the complete schedule is infeasible given the resource constraints. One way of restricting the number of alternative actions is to set a threshold value for the maximum difference in expected costs between the optimum action and the alternatives, chosen by consideration of the likely sensitivity of the cost of any one inspection to the total IRM budget.

CS Scheduling This produces a usable schedule of IRM actions for the complete structure, by combining inspection plans for each joint and generating new schedules. The procedure is described in the next section.

3.3 Generating the Final Schedule

The final constrained schedule takes into account global resource constraints. The initial schedule is given by the locally optimum actions and has to be modified to take into account resource constraints such as global safety requirements, length of time available for inspection and number of divers, available inspection equipment, location of diving vessel, and guidelines and recommendations.

3.3.1 *Modelling Scheduling Data*

The quality of a schedule may be measured by two related factors: cost and satisfaction of resource constraints. Since the cost of inspection and repairing for each action has already been evaluated, an estimate for the total cost for each schedule is simply estimated as the sum of the individual costs. This cost does not include the added costs of exceeding any scheduling constraints. As there is no direct method of evaluating the exact overall cost for the complete schedule, the satisfaction of resource constraints, measured by *resource deficits* (RD), is considered instead. For a schedule to satisfy completely all constraints requires that all RDs be positive or zero for all inspection periods.

To take into account more than one resource, a schedule utility function could be defined as a combination of weighted resource parameters to enable automatic comparisons between schedules. In practice operators prefer the user to carry out closer consideration of schedules to make a final decision.

Initially the System will only consider CS scheduling if any inspection period has too many actions assigned to it and the total resources required for all actions to be carried out is less than or equal to the total amount of available resources over the scheduling period. At the end of the CS Scheduling procedure, a schedule with no (if possible, if not, then fewer) inspection periods with positive resource deficits is output. The output is in essence a very simple linear schedule, in that a window of time is allocated as a resource and a list of inspection tasks for a year could be assumed to be a sequence of activities, albeit unordered.

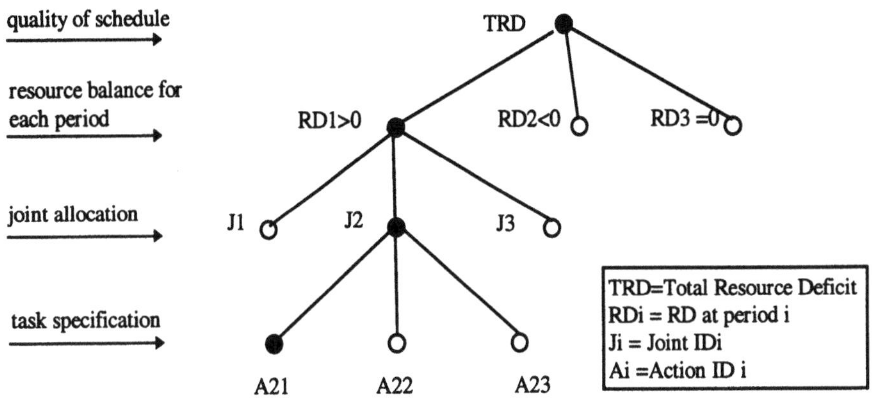

Figure 3 Constraints satisfaction scheduling tree

The resource deficit information is embedded into an *AND/OR tree* representation of the scheduling objects. The root node indicates the total resource deficit. The scheduling procedure decomposes the problem to allow parts of the schedule to be considered in turn and hence the next two levels are AND nodes. The first nodes below the root represent action periods and each has an associated resource deficit. Below each of these are the nodes representing the joints to be inspected. Each component in turn has a default action as well as a number of alternative actions and these are the leaf OR nodes. The task of generating a practical plan then becomes a problem of systematically searching for alternative actions that reduce or eliminate positive resource deficits.

3.3.2 Searching for the Schedule

Brute force depth- or width-first search methods choose the first alternative action encountered satisfying the goal, i.e., reducing resource deficits. A *heuristic search* strategy is required that allows selection of the next best solution.

```
function heuristic_based_search_method ()
{for ?period inlist all instanceof schedule;
  do { ?k = 1;
  while{?k<schedule.total_alternatives, ?period.resource_deficit>0;}
  do { for ?joint inlist all ?period.joints;
       do {  focus_on_next_action(?joint);
             ?k = ?k + 1;}
       for ?joint inlist all ?period.joints;
       do {sort_by_slot(?period.joints, ?joint.cost);
          for ?joints inlist all ?period.joints;
          do{if ?joint.current_action_no<
                                              ?joint.total_alternatives;
             then {?action = ListNth(all ?joint.actions,
                                              ?joint.current_action_no);
             if test_candidate_action(?period,?joint,?action = TRUE;
             then update_schedule(?period,?joint,?action);}}}
}}}
```

Figure 4 Heuristics-based Searching Function

Figure 4 shows a cost check incorporated into the search. In this figure, a variable *?fred* represents an instance of an object, and *?fred.foo* refers to its slot *foo*. Function *focus_on_next_action* forces the search along the leaves of the tree; *test_candidate_action* executes a set of rules to check the constraints on the tested action; *update_schedule* is the constraint propagation procedure used to update the schedule and to recalculate resource deficits. The *sort_by_slot* step ensures that the next cheapest action is selected.

At the end of the searching procedure, the user has a selection of feasible schedules from which to choose a final schedule. Changing resource constraint settings allows what-if analysis. It is convenient at this point to consider if other joints should also be included usefully in those inspection periods without a resource deficit. These additional joints may be *neighbouring*, and hence it is practical to include them, or they may be *similar* joints to those already included for inspection, so should also be of concern. At the end of this process, a rational combined schedule is selected by the user.

4. Discussion and Conclusions

Structural reliability analysis requires large amounts of data to be assembled, guidance during execution of modules and intelligent results interpretation. Risk analysis, taking costs into account, is even more knowledge intensive. The costs of inspection can be high: for instance the cost of employing more than one vessel (from which only two divers can operate) from which the inspection operations are carried out or moving this around the structure are prohibitively high (£50000-£100000 per day). Thus ensuring that inspections are restricted in numbers and technique and region of the structure whilst maintaining reliability would be a great advantage. There is also a trade-off between the employment of high-performance and efficient inspection teams and vessels that is difficult to evaluate without making use of a system such the one described in this paper.

A similar KBS approach to support reliability-based IRM planning would be suitable for other structures such as air-frames and steel bridges, and more specifically, the jack-up type platform with very similar IRM planning issues, and pipelines. Other damage modes and the intelligent support for data modelling of uncertainties, embedding human factors, for instance, is also required. A further extension would be to consider IRM planning for a field of platforms or a group of related structures where resources are often shared.

The aim of this paper was to present a framework for use of AI-based searching techniques to provide intelligent support for reliability-based IRM planning. A demonstrator was developed to illustrate the architecture and heuristics and was tested against an example structure. Other applied AI techniques, such as genetic algorithms for generation of new IRM schedules, case-based reasoning for standard cases of IRM planning and the use of neural nets and fuzzy logic for data input, could also be investigated for integration with structural reliability analysis.

References

[1] Thoft-Christensen, P, Baker, MJ, (1982) Structural Reliability Theory and its Applications, Springer-Verlag

[2] Peers, SMC, Dharmavasan, S, Kam, JCP, et al, (1992) A Rational Inspection Scheduling Philosophy for Fixed Offshore Structures, Proc 11th Ocean, Marine and Arctic Engineering Symposium, OMAE 92, Calgary, Canada

[3] Faber, MH, Dharmavasan, S, Dijkstra, O, (1994) Integrated Analysis Methodology for Reassessment and Maintenance of Offshore Structures, Proc of 13th Ocean, Marine and Arctic Engineering Symposium, OMAE 94, Houston, Texas, US

[4] Yang, Q, (1997) Intelligent Planning - A Decomposition and Abstraction Based Approach, Springer

[5] Berry, PM, (1992) The PCP: A Predictive Model for Satisfying Conflicting Objectives in Scheduling Problems, Artificial Intelligence in Engineering, 7, pp 227-242

[6] Duda, R., Gaschnig, J., Hart, P., (1979) Model Design in the PROSPECTOR Consultant System for Mineral Exploration, Expert Systems for the Microelectronic Age, ed. D. Michie, Edinburgh University Press

[7] Guilfoyle, C, (1986) Ten Minutes to Lay, Expert System User, August

[8] Apte, CV, Weiss, SM, (1987) An Expert System Methodology for Control and Interpretation of Applications Software, Int. Journal of Expert Systems, 1(1), pp17-37

[9] Katsoulakos, PS, Hornsby, CPW, (1988) Expert Systems and Marine Applications, Cooloquium, The Institute of Marine Engineers, October 18

[10] Willoughby, AA, Laures, J-P, (1990) Development of a Knowledge-Based System for Assessing Flaws in Welded Structures, Proc of Expert Systems 90 Conference, September

[11] McMahon, CA, Banerjee, S, Sims Williams, JH, et al, (1994) Hypertext and Expert Systems Application in Fatigue Assessment and Advice, Automation in Fatigue and Fracture: Testing and Analysis, ASTM STP1231, ed. C. Amzallug, Philadelphia

[12] Khong, VH, Lucia, AC, (1990) RAMINO - A Knowledge Based System and a Co-ordinator of Numerical Models for Structural Reliability Assessment, Proc 10th Int Workshop on Expert Systems and their Applications, Avignon

[13] Ahmad, K, Langdon, A, Frieze, PA, (1991) An Expert System for Offshore Structure Inspection and Maintenance, Computers and Structures, 40(1), pp 143-159

[14] Besse, P, Seridji, A, Pelletier, JL, et al (1992) Advanced Risk Management System (ARMS) A Resource Constrained Normative Decision System for Offshore Platforms, Oil and Gas in a Wider Europe, Proc 4th EC Symposium, Berlin

[15] Chen, K, Zhang, S, Huang, W, (1996) Artificial Intelligence Beta -unzipping Method in Structural System Reliability Analysis, Computers & Structures, 60(4), pp 619-626

[16] Auflick, JL, Hahn, HA, Morzinski, JA, (1998) HuRA! A prototype expert system for quantifying human errors, Proc of 4th Int Cnf on Probabilistic Safety Assessment and Management PSAM4, Springer

[17] Le Pape, C, (1993) Classification of Scheduling Problems and Selection of Constraint-Based Techniques, Colloquium on Advanced Software Technologies for Scheduling, Institute of Electrical Engineers, London, UK, 26 April

[18] Tate, A,Hendler, J, Drummond, M, (1985) A Review of Planning Techniques, Knowledge Engineering Review, 1(2), pp4-17

[19] Thornton, C, Du Boulay, B, (1992) Artificial Intelligence Through Search, Intellect Books

[20] Miyashita, K, (1995) Case-based Knowledge Acquisition for Schedule Optimisation, Artificial Intelligence in Engineering, 9, pp27-287

[21] Duchessi, P, O'Keefe, R, (1990) A Knowledge-based Approach to Production Planning, Journal of Operational Research, 41(5), pp337-390

[22] Bolc, L, Cytowski, J, (1992) Search Methods for Artificial Intelligence, London: Academic Press

[23] Atabaksh, H, (1991) A Survey of Constraint Based Scheduling Systems Using an Artificial Intelligence, Artificial Intelligence in Engineering, 6(2)

[24] Stefik, M, (1981) Planning with Constraints (MOLGEN Part 1), Artificial Intelligence, 16(2)

[25] Kumar, V, (1992) Algorithms for Constraint Satisfaction Problems: A Survey, AI Magazine, Vol 13(1), pp32-44

TIGER with Model Based Diagnosis: Initial Deployment

Dr. Robert Milne and Dr. Charlie Nicol

Intelligent Applications Ltd, 1 Michaelson Square, Livingston, West Lothian. Scotland. EH54 7DP

Dr. Louise Travé-Massuyès

Intelligent Control Systems Group, LAAS-CNRS, 7 Ave du Colonel Roche, 31077 Toulouse, France

Abstract. The TIGER system is in continuous use on over 20 gas turbines across four continents, performing fault diagnosis of industrial gas turbines used for power generation. TIGER receives several hundred data inputs at a once per second interval and performs several layers of fault detection and diagnosis using expert system techniques in a temporal reasoning rulebase.

Development of the system has now extended for more than six years and more than 80 man-years. The diagnostics are supported by an extensive user interface, data archive, trending and display system. Across the 20+ current locations, TIGER has helped to identify and diagnose over 700 incidents representing 200 different fault categories. Because of the high value of gas turbines and the high expenses in operating them, TIGER's benefit is considerable and has been documented at $150,000 or more for most of the locations.

This paper gives a summary as TIGER is today, the benefits to many of the sites and an overview of the recently added model based diagnosis module, Ca~En.

1. INTRODUCTION

Industrial gas turbines are becoming the workhorse of the power generation industry. New power plants can be built relatively quickly and cheaply, and when all goes well, they run continuously for long periods of time with very little attention required from operators. However, in the increasingly competitive electricity generation industry, profit margins are very tight, requiring good return on investment and small margin for the cost of unexpected maintenance problems or failures. In addition, most sites use the minimal level of manning that is practical and there is an increasing skill gap between the complexities of a gas turbine and the things that can go wrong, and the operators in charge of ensuring that the machine runs day-to-day.

In fact, a major trend is to run the gas turbines unattended with only remote support. Hence, in order to make a good profit for a power generation company, it is vital that the gas turbine has high availability, minimum unexpected maintenance costs, and at the same time, can be operated by someone who has relatively little skill and training in the intricacies of a gas turbine.

What is needed is a system that can help the plant increase its availability by avoiding shutdown time, either as an unplanned failure of the system or if a failure does occur, minimising the time that the gas turbine is down. To achieve this the goal is condition monitoring [1]. Assessing the condition of the gas turbine continuously and detecting at the earliest possible stage any developing faults or problems. When a problem has been detected it is also vital to diagnose it as completely as possible, and as quickly as possible.

Because the detailed knowledge of how a gas turbine operates and problems manifest themselves what is needed is a system which can make good diagnostic knowledge and information available to the people responsible to ensure the gas turbine is available when needed [2].

The aim is to detect problems at the earliest possible stage and be able to rectify them before they become expensive failures. If an unexpected failure does occur, then to rapidly understand what the problem was so that it can be fixed as efficiently as possible, and the machine returned back to power generation quickly. Another important factor is to get optimum efficiency out of the machine since machines burn $1.5 million worth of gas every month, 1-2% drop in efficiency can be very expensive.

The TIGER software was developed to address these requirements. It receives 600 measurements from the gas turbine controller at once per second intervals and performs an extensive fault detection and diagnosis every second. TIGER's task is to check the data inputs just as the best gas turbine engineer would, and ensure that everything is operating at the correct level. More importantly, as it checks dynamic changes such as an increase in power level, to confirm that the gas turbine has reacted properly in doing this.

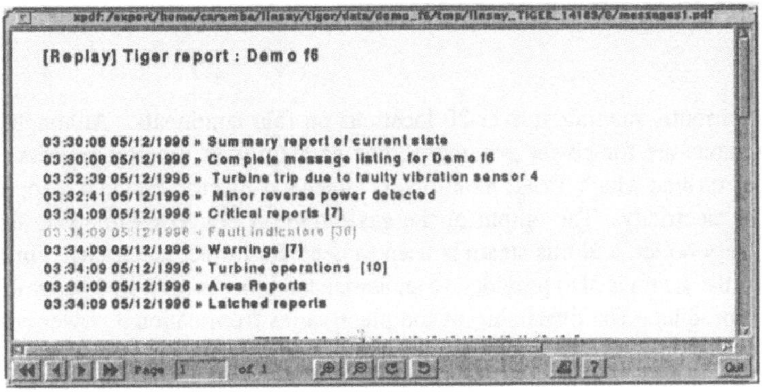

Figure 1: The TIGER fault manager top level

It is impossible to diagnose all faults which occur on a gas turbine, so TIGER's diagnostic system is structured in a layered way. At the lowest level is an extensive fault detection set of mechanisms which look for a wide variety of abnormalities. These are then combined using a patented temporal rulebase system. A process of hierarchical abstraction consolidates conclusions with high confidence until the system is able to reach a diagnostic conclusion for a particular event. Very often the cause of a problem is not known to TIGER from its data inputs or possibly not known at all to gas turbine engineers. In this case TIGER reports a partially consolidated conclusion to assist the engineer in focussing on the problem without actually diagnosing the exact cause. This strategy of always trying to provide a helpful focus for a problem greatly enhances its usefulness.

The TIGER diagnostics are presented through a fault manager using hypertext. The most important problems are presented at the highest level and the hierarchical tree leading to the fault conclusions, can be viewed by exploration with a mouse. In fact, the output of the fault detection and diagnosis levels are presented three times: first in a complete time sequence; second graded by priority from critical, warning, detected incidents, or operational incidents; and finally, grouped by the sub-systems of the gas turbines.

To allow a gas turbine engineer to further troubleshoot problems, TIGER has an extensive set of displays, graphical and trending tools. The graphical system is used to display the relevant data inputs in a close time sequence to verify the conclusions of the diagnostic system. This has proven invaluable for permitting engineers to have confidence in a diagnosis.

The trend system automatically builds a 24 hour trend for all 600 inputs, each day. Weekly and monthly trends can be constructed easily. This provides not only a way to check for long term deterioration, but also to get a rapid overview of the data.

2. TIGER IN USE TODAY

TIGER is currently running at over 20 locations on four continents. Although most of these locations are for power generation, one of them is in use at a process plant. A typical gas turbine which TIGER monitors, is a General Electric Frame 6 generating 40 Mwatts of electricity. The output of the gas turbine is converted to steam through a heat recovery boiler, and this steam is used to generate further electricity. In many of the plants, the steam is also provided to an associated process such as a paper mill as an important product. The ownership of the plant varies from national power companies to small independent power producers.

TIGER is also in use on four oil platforms in the North Sea providing power generation on the oil platforms for major oil companies. In approximately half of the locations, TIGER can be accessed remotely so that a central monitoring office in Scotland is able to check each gas turbine on a daily basis and also provide remote diagnostics support. Companies using TIGER include: National Power, Powergen, Exxon Chemical, British Petroleum, Chevron, Oryx, and Kvaerner Energy.

There are two main groups of users for TIGER. At the plant level an operator can check TIGER every shift and see if any important warnings are being presented. It is also used locally when a problem does occur. Generally, the operators don't have in depth knowledge of the gas turbine and are only concerned with faults requiring immediate action. The second group, and most common user type, is the engineer responsible for the operation of the gas turbine. The engineer needs to know about developing problems and will also troubleshoot more difficult situations. The range of problems detected by TIGER is most valuable to this type of engineer, hence giving him a high motivation to use the tool.

For locations where there are not engineers on site, the TIGER system is monitored and acted upon remotely by the TIGER monitoring team in Scotland. This remote technical backup has considerable benefit in saving travel costs and time, particularly for offshore oil platforms.

At every installation, TIGER has been invaluable in identifying a wide range of faults and problems which were previously undetected. It has now been used in a wide variety of situations to further troubleshoot and identify the cause of any problem so that rectification can be made.

3. TIGER DIAGNOSTICS

The TIGER system is a layered diagnostic system working at increasingly higher levels of abstraction. The lowest levels are a wide variety of specialist fault detection mechanisms. They are originally implemented in an expert system environment but have been coded directly into C++ for high efficiency. Many of the fault detection checkers are tuned to the performance of a particular gas turbine through an automatic tuning algorithm that examines one month's worth of data before setting the patterns of normal behaviour.

The fault detection mechanisms check that parameters are within normal operating limits for that gas turbine, but also look at rates of change and relationships to other variables. The output of the fault detection level is processed by a second level rulebase which uses time constraints between events to abstract events into actions on the turbine. Although it can be viewed as a classical knowledge base, it now has several thousand diagnostics based on the wide variety of incidents. These diagnostics are implemented in an expert system environment and then compiled to C++ code for faster execution.

Layer 4: Diagnostic Conclusions:
sorted by: time sequence, priority, sub-system

Layer 3: High level rules - patented temporal reasoning system

Layer 2: Hierarchical aggregation of detected events

Layer 1: Fault detection - extensive set of special purpose detectors

Figure 2: The layers of TIGER diagnostics

At the upper levels, a set of high level rules consolidate a combination of detected events and lower level rule conclusions into higher level conclusions. These might be a consolidation of faults or it might be a specific diagnosis such as a trip caused by faulty sensors. The approach to combine these is essentially a rulebased temporal reasoning system, but using a patented mechanism.

The TIGER diagnostic system levels of hierarchy are summarised as follows.

Fault Detection

The purpose of this level is to detect any abnormalities or unusual events on the gas turbine. Currently within the TIGER system, there are over 4000 fault detectors checked every second. These vary from simple context sensitive limit checkers to specialised mechanisms to ensure the dynamic response of the turbine is as it should be.

Diagnostic Rules

Using well established forward chaining techniques from knowledge based systems, TIGER uses diagnostic rules to specify the correct interaction of parameters and sub-systems within TIGER, and determines whether they are functioning properly or not. This provides a very sophisticated ability to detect whether the working of the turbine is correct. This mechanism uses temporal reasoning techniques to track sequences of events over time.

High Level Rules

The TIGER fault manager uses a patented [3] temporal reasoning approach to combine the results of the fault detection and diagnostic rule levels into higher level conclusions. In essence, this module is a forward chaining rule base. The temporal techniques allow it to collect related events in a time window. This resolves the problem that related events aren't always detected at the same time. For example, when there is a time window on a fault detector or a valve takes a few seconds to move. TIGER combines the thousands of diagnostic checks it performs in a hierarchical fashion so that the user sees a single fault conclusion for a major incident.

Accounts For

Once a problem has been identified, the existence of this problem accounts for many other diagnostic messages. TIGER organises the wide range of phenomena that occurs when a problem exists through the 'accounts for' set, so that a single conclusion can account for many consequent results. This organises the many events, very simply for the user, and reduces the alarm explosion that can occur during a major incident.

Areas

When it is not possible to provide a concise diagnosis, the fault detection and low level diagnostic conclusions are grouped by sub-systems or 'areas', such as, all the messages related to the fuel system, the inlet guide vanes or the combustion system. This allows an experienced engineer to rapidly focus on the core problem area and gives him support in determining what problem has just occurred.

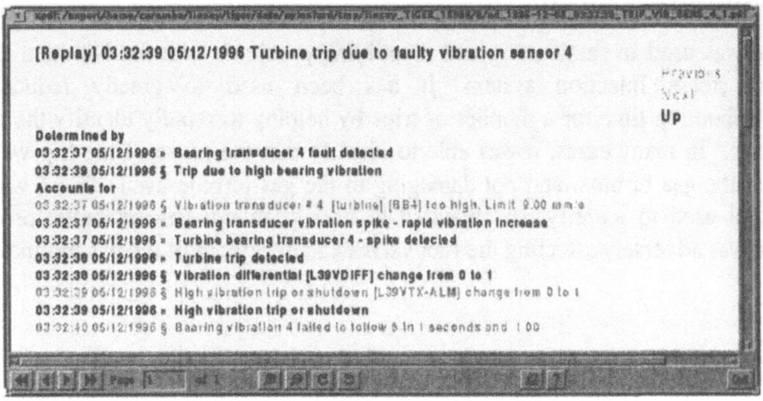

Figure 3: A TIGER rule display

The diagnostics were built both 'bottom up' and 'top down'. For each sub-system of the gas turbine, a failure analysis was conducted identifying how the system should work, what might go wrong, and how that would be indicated by the TIGER variables. This led to a ground up development of a wide variety of rules. Currently the main driver of new diagnostics, is by examining incidents which actually occur and creating improvements or new diagnostic rules. A knowledge engineer carefully examines each incident which has been logged in the TIGER incident database, checks what diagnostic conclusions TIGER reach and then makes any necessary improvements or adjustments.

Although gas turbines are relatively well understood, it is incredible how much they get up to in unexpected ways. Many strange behaviours and actions occur which engineers are surprised with and do not understand immediately, hence, being data driven has been very productive to the improvement of the TIGER diagnostics. This data driven approach has been driven by over 700 incidents logged in the TIGER incident database, and approximately 25 fired years of gas turbine operation.

From the Artificial Intelligence viewpoint, these are relatively straight forward techniques and technical approaches. However, the complexity of the system requires considerable experience to develop and evolve.

4. TIGER BENEFITS

The following is a summary of locations where TIGER has produced a saving of at least $150,000. This represents the majority of the installations which have been in place for at least one year.

Aylesford Newsprint

TIGER was used in cataloguing and identifying problems with the Nitrogen Oxide (NOx) steam injection system. It has been used to greatly reduce the troubleshooting time for a number of trips by helping to rapidly identify the cause of a trip. In many cases, it was able to identify that the cause of the trip was not part of the gas turbine, and not damaging to the gas turbine itself. TIGER was the key tool used to identify hat appeared to be a problem with hydraulic pressure which was adversely affecting the fuel valves and behaviour of the gas turbine.

BASF

TIGER helped to identify and correct a number of initial commissioning and installation problems. One of these was a problem with the stability of the steam supply for NOx injection. Based on the documented behaviour of the gas turbine through TIGER, a refurbishment of the steam injection supply system was conducted which eliminated many of the problems.

Alba Platform

TIGER was instrumental in identifying how liquid fuel start problems were because of instability of the liquid fuel system at low load. This was causing problems with starting the gas turbines on liquid fuel and directly affecting their availability.

Dubal

TIGER identified over 30 problems that were affecting the turbines behaviour and operation on a regular basis, but which were unknown prior to its installation. In addition, there were two failures of the atomising air compressor, which were rapidly diagnosed using TIGER and helped to considerably reduce the downtime.

Exxon Chemical

TIGER has become the primary tool for the engineers to monitor and analyse the behaviour of the gas turbine. This then provided critical data in the de-bottlenecking process for the plant. A major bottleneck was the reaction of the second stage nozzles at high power. TIGER was used to documentation and find a resolution to this problem.

At higher power levels, the problem again reoccurred and TIGER was used to understand and resolve the issues. As Exxon extracted higher power out of the machine, the window into the gas turbine, provided by TIGER, was critical in supporting this process and enabling them to get more power out of the turbine and a higher output out of the plant.

Fellside

Prior to the installation of TIGER it was noted that first stage nozzles were having to be changed more frequently than expected. Almost immediately after TIGER was installed, the problem was identified as major oscillations caused by the fuel control system. TIGER was instrumental in the troubleshooting and fault isolation process. Using the knowledge gained through the TIGER analysis, the problem was rectified with considerable savings.

Hjorring

TIGER was able to identify a number of intermittent problems with the valve movements and flame scanners, as well as the DLN1 NOx control system. The experienced site engineers used TIGER to track down a range of minor problems and faults affecting the behaviour of the gas turbine and improve its operation.

Kemsley Mill

Due to a developing problem in the combustion system, the turbine was operating with a very high exhaust spread, however, it was not convenient to shut the turbine down to rectify it. Using TIGER as a window into the behaviour of the gas turbine, the site was able to nurse the gas turbine operation through to a weekend when they were able to have the appropriate parts and maintenance engineers ready to correct the problem as quickly as possible.

Through TIGER they were able to get the maximum output of the turbine without tripping it through the period until it was convenient to shutdown. By delaying the shutdown until they had all the equipment and personnel ready, plus at a lower tariff time, it provided considerable savings to the site.

ETAPS Turbine 1 & 2

On this offshore oil platform, neither gas turbine was able to start on liquid fuel and gas fuel was not available. An engineer used TIGER to rapidly identify the precise failure within the liquid fuel system and direct the repair work as efficiently as possible. After expensive downtime, by using TIGER, the site was able to get the gas turbines up and running quickly, providing considerable benefit on the oil platform.

TIGER was also used to catalogue all the problems and deficiencies from the initial installation of the gas turbine. This enabled the customer and the current holder of the maintenance contract to make a warranty claim against the initial installation. TIGER documented all the problems affecting the behaviour, efficiency and operation of the gas turbine at a very early stage so that they could be rectified before they caused further long-term problems.

TIGER has consistently identified a number of faults and problems on gas turbines including many of which were not known, but their rectification will lead to a more efficient and smoother long-term operation.

5. TIGER DEVELOPMENT

The first prototype of the TIGER system was built with Exxon Chemical in Scotland, approximately seven years ago. This then led to a three year ESPRIT supported project to develop the initial TIGER prototype [4], [5], [6]. That project involved a collaboration of Intelligent Applications Ltd, Exxon Chemical Olefins Inc, John Brown Engineering, Dassault Aviation, CNRS-LAAS, and University Polytechnic of Catalonia.

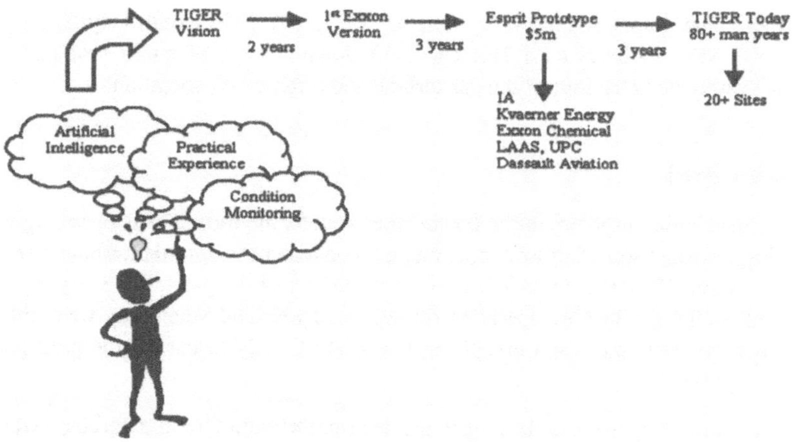

Figure 4: The development of TIGER

After the ESPRIT project, John Brown Engineering and Intelligent Applications worked together to commercialise the system and have continued to further develop the software and diagnostics.

A key aspect of the development team is that it combines artificial intelligence developers, gas turbine experts and software engineers. Each provides expertise in a complementary area.

At this point, the investment in TIGER has continued over a six year period and totals more than 80 man years and $5 million of development. Development is still continuing at a rapid pace. Future developments include continued expansion of the diagnostics, widening the range of gas turbines for which TIGER is available, enhancing the performance monitoring functionality and the addition of model based prediction and isolation techniques.

Each TIGER installation is 'almost' standard. TIGER has to be configured to the exact sensors which are installed on each gas turbine. Although these are very similar, the detailed differences have to be identified and the configuration adjusted. TIGER will then auto tune its fault detection systems to the way a particular gas turbine operates.

The diagnostics automatically activate (or deactivate) depending on which inputs are available. The configuration for a TIGER for a Frame 6 gas turbine takes about two days. Initial user training takes about one day, reflecting the ease of use of the system.

6. MODEL BASED FAULT DETECTION AND DIAGNOSIS

Although the core of TIGER is a temporal rulebased system, there is an important class of faults to detect which require a model based approach, that is, the predicted value for a measurement depends on a complex combination of several inputs. One area where this is particularly important is predicting the expected power output of the gas turbine, giving a combination of atmospheric air conditions, fuel setting, and steam injected for pollution control.

The focus of current development is to use a qualitative interval predictor based on the Ca~En software [7], [8]. (Ca~En is short for Causal Engine) This combines a model based fault detection ability, by predicting what expected outputs should be, with a model based fault isolation mechanism providing this additional leverage over rulebased implementations [9]. The initial integration of Ca~En with Tiger is described in [6]. The recent work has been to re-implement the models and provide a much tighter integration with Tiger than was included in the original Tiger ESPRIT project. Qualitative modelling is used to generate the expected value of key measurements, such as a fuel valve position.

The qualitative predicted predictions system is being used on critical sub-systems of the gas turbine such as the gas and liquid fuel systems.

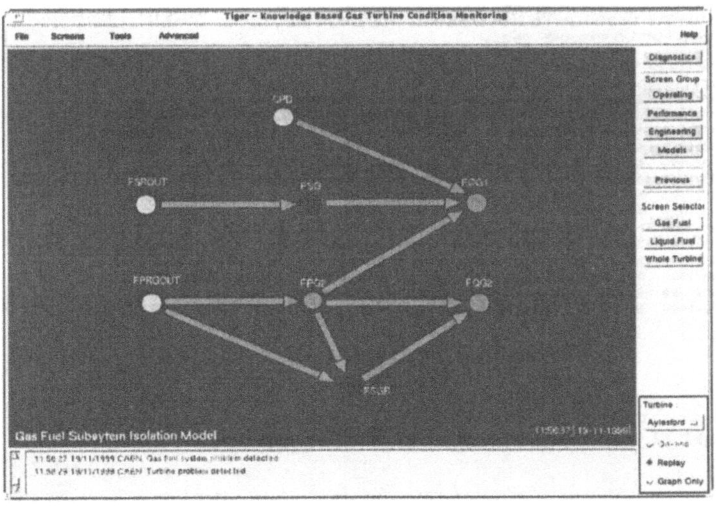

Figure 5: Ca~En Activation Graph of Gas Fuel System

The combination of TIGER with Ca~En provides a much more sensitive fault detection capability for the gas fuel and other sub-systems. In addition, it's built in fault isolation mechanism models the physical operation of the gas fuel system and its detailed components to identify which are working properly and which are not. This has provided much better accuracy of fault detection on the "spike problem" which existed at the customer site.

The integration of Ca~En with TIGER is at several user interface levels. The top-level display shows the activation conditions and inputs to the model. The screen snapshot in Figure 5 shows the activation display of Ca~En when a fault is detected. The variables used within the models are highlighted as nodes on the display:

> green meaning 'not detected as faulty'
> red meaning 'detected as faulty' or 'not okay'.

The arcs between the nodes represent the components that provide the translation of the input of a component to its output. Note, that because of the resolution of the sensors available, very often there are several components and sub-systems, and possibly the actions of the controller on each arc.

The diagnostics from the modelling system are fully integrated with the output of the other TIGER diagnostic modules, in the above situation, the diagnostic messages are as follows. Note also the fault detection messages from the envelope predictions.

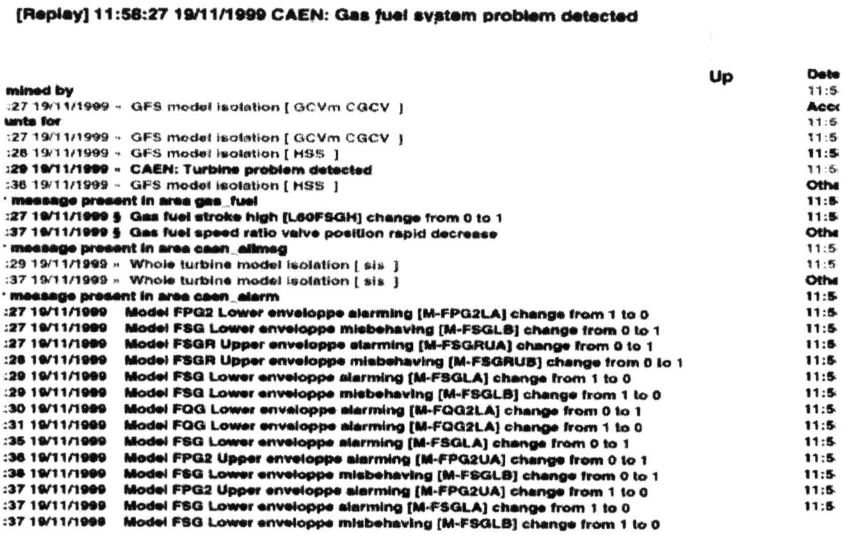

FIGURE 6: TIGER Fault Manager Screen

The output of Ca~En is integrated with the fault manager.

These messages are also then used in the high level rules of TIGER which are fully integrated as part of the standard TIGER diagnostic output. When looking at a high-level rule conclusion the messages from Ca~En are colour coded brown so that it is easy to visually see the contribution of the models to the high level rules.

The Ca~En diagnostics often produce one or more sets of possible faulty components, these are displayed by clicking on the diagnostic message within the fault manager. For each sub-system, its schematic is then displayed, highlighting the possible faulty components.

The specific Ca~En diagnostics can be examined, Figure 7 showing the gas fuel sub-system with the relevant faulty component highlighted, as well as the possible sets of faulty components. This allows the engineer to rapidly understand the conclusions reached by the model based system.

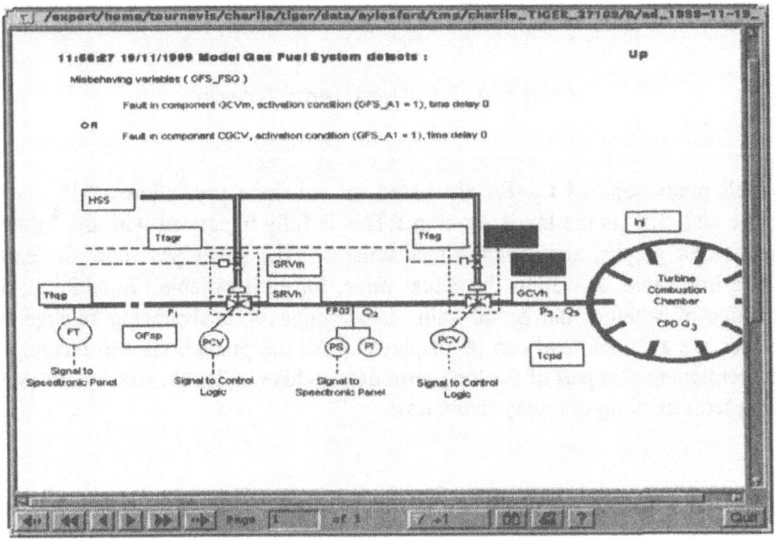

Figure 7: Fault Manager of Ca~En Diagnosis

All this is displayed on the model based predictions from Ca~En as illustrated on the following graph. The channels with the prefix "M-" are from the models.

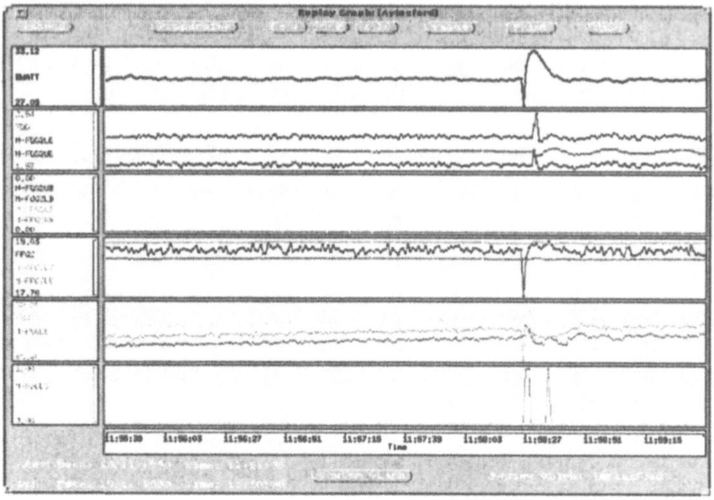

Figure 8: Incident Graph

The fault predictions of Ca~En are based on envelope predictions. "UE" is upper envelope and "LE" is the lower envelope. This is fully integrated with the TIGER data stream, TIGER graphs, and trending mechanisms. The envelopes show the expected high and low value, as well as the actual value, for each variable. In addition, digital indications of whether the actual value is alarming or misbehaving relative to the envelopes are available and can be displayed. All the predictions and envelopes are stored permanently as part of the long-term data archive in TIGER, so they are available for long-term trending or replay at any time.

Testing of Ca~En

Since the beginning of this year we have been carrying out extensive testing of the TIGER /Ca~En system. Over 25-batch test runs were conducted varying from 2 weeks to 2 months at a time covering key periods of operation for the gas turbine. In addition, the combined system has been running continuously at the site since April 2000.

Poor Response of Liquid Fuel Valve

As an example of the benefit of Ca~En, consider a liquid fuel problem. This is a significant problem at the site, and therefor it was important that Ca~En was able to detect and isolate the fault. The response of the valve to the set-point appears to be deteriorating over time, and could eventually prevent start-up on liquid fuel. Based on the results tabulated below, we can see that the fault detection capability of Ca~En was good, detecting and isolating the faults in every case, whereas TIGER does not detect the fault. At this stage it is not known what the cause of this fault is, although it may be related to other problems with the liquid fuel system. Interestingly, the TIGER diagnostics do not detect this because its limits must be too wide in order to allow for the wide range of normal operation. Hence, this is a clear example of where the model based system is detecting and isolating a fault that could not be otherwise.

Incident No.	Fault	Ca~En Fault Detect %	Ca~En Fault Isolate %	Ca~En False Alarms	Tiger Detect %	Tiger Diagnose %	Tiger False Alarms
10.1	Fuel Valve Lagging Behind Set Point by 1%	100	100	0	0	0	0
10.2	Fuel Valve Lagging Behind Set Point by 1%	100	100	0	0	0	0
10.3	Fuel Valve Lagging Behind Set Point by 1%	100	100	0	0	0	0
11.1	Fuel Valve Lagging Behind Set Point by 1%	100	100	0	0	0	0
26.1	Fuel Valve Lagging Behind Set Point by 3%	100	100	0	0	0	0

Table 1: Ca~En test results

Although Ca~En add a valuable capability to Tiger, the model based system's power is restricted primarily by the availability of key measurements to allow it to isolate faults with high resolution. Unfortunately the measurements of a gas turbine are generally provided to support monitoring that it is working safely rather than diagnosing and isolating faults.

7. SUMMARY

This paper has presented the TIGER knowledge based condition monitoring system for gas turbines. It is in use at over 20 locations providing continuous fault detection and diagnosis for gas turbines used in power generation. TIGER has consistently found a wide variety of faults and problems for gas turbines and has produced benefits in excess of $150,000 at all the locations for which it has been installed for more than one year.

The system uses a temporal reasoning knowledge based system and was developed with classic knowledge acquisition techniques primarily driven by real examples from gas turbines. Recently, a model based diagnosis module, Ca~En, was added. It is interesting that the AI techniques used are not complex and innovative. Instead, well-established forward chaining rule based techniques have been demonstrated to be adequate for most requirements. The power is the knowledge, not the technical approach. The system continues to be developed and is being installed on increasing numbers of gas turbines.

8. REFERENCES

[1] Handbook of Condition Monitoring - Techniques and Methodology. Chapter 22, 521-540. Editor: A. Davies. Publishers: Chapman & Hall (1998).

[2] R. Milne, Knowledge Based Systems for Condition Monitoring, *Handbook of Condition Monitoring - Techniques and Methodology*. Chapter 22, 521-540. Editor: A. Davies. Publishers: Chapman & Hall (1998).

[3] United Kingdom Patent Application No: GB2323197A. Filed in 1997 by applicants Intelligent Applications Ltd.

[4] R. Milne, C. Nicol, M. Ghallab, L. Trave-Massuyes, K. Bousson, C. Dousson, J. Quevedo, J. Aguilar and A. Guasch, Real-Time Situation Assessment of Dynamic Systems, *Intelligent Systems Engineering Journal*. Autumn, 103-124 (1994).

[5] R. Milne, C. Nicol, L. Trave-Massuyes, J. Quevedo, TIGER: Knowledge Based Gas Turbine Condition Monitoring, *AI Communications Journal.* Vol 9, N° 3, 1-17. Publishers: IOS Press (1996).

[6] L. Trave-Massuyes and R. Milne, Gas Turbine Condition Monitoring Using Qualitative Model Based Diagnosis, *IEEE Expert magazine, "Intelligent Systems & Their Applications*". Vol 12, N° 3, 21-31. Publishers: IEEE Computer Society (1997).

[7] K. Bousson and L. Trave-Massuyes, Fuzzy Causal Simulation in Process Engineering, *Proceedings of IJCAI-93* (1993).

[8] K. Bousson and L. Trave-Massuyes, Putting More Numbers in the Qualitative Simulator Ca~En, *Proceedings Second International Conference Intelligent Systems Engineering*, IEE Publications, Stevenage, UK, 62-69 (1994).

[9] L. Trave-Massuyes and R. Milne, Diagnosis of Dynamic Systems Based on Explicit and Implicit Behavioural Models: An Application to Gas Turbines in ESPRIT Project TIGER, *Applied Artificial Intelligence*. Vol 10, N° 3, 257-277. Publishers: Taylor & Francis (1996).

TIGER is a Trademark of Intelligent Applications Ltd.

AUTHOR INDEX

GPSR Compliance

The European Union's (EU) General Product Safety Regulation (GPSR) is a set of rules that requires consumer products to be safe and our obligations to ensure this.

If you have any concerns about our products, you can contact us on ProductSafety@springernature.com

In case Publisher is established outside the EU, the EU authorized representative is:

Springer Nature Customer Service Center GmbH
Europaplatz 3
69115 Heidelberg, Germany

Batch number: 10682323

Printed by Printforce, the Netherlands